Mathematical Marvels: Texts and Monographs in the Spirit of CR Rao

This series aims to publish books in all areas influenced by Dr. C.R. Rao, including statistics, linear algebra, econometrics and information geometry. It also includes content in other domains of mathematical sciences, applied statistics, financial mathematics, operations research and computer science with primary focus on mathematics and statistics. The series aims to carry on the tradition of C.R. Rao's innovative and influential works, which have impacted a variety of fields, and to provide top-notch textbooks and monographs in related areas.

Bipan Hazarika • Hemanta Kalita

Theory of Henstock–Orlicz Spaces

Springer

Bipan Hazarika
Department of Mathematics
Gauhati University
Guwahati, Assam, India

Hemanta Kalita
Department of Mathematics
VIT Bhopal
Bhopal, Madhya Pradesh, India

ISSN 3005-1967 ISSN 3005-1975 (electronic)
Mathematical Marvels: Texts and Monographs in the Spirit of CR Rao
ISBN 978-981-96-9547-8 ISBN 978-981-96-9548-5 (eBook)
https://doi.org/10.1007/978-981-96-9548-5

Mathematics Subject Classification: 46B03, 46B20, 26A39, 46G10, 46A39, 46E30

This Springer imprint is published by the registered company Springer Nature Singapore Pte Ltd.
The registered company address is: 152 Beach Road, #21-01/04 Gateway East, Singapore 189721, Singapore

Preface

The analysis of function spaces, particularly Orlicz spaces, defined on arbitrary measure spaces involves understanding their structure and properties. Orlicz spaces are generalizations of Lebesgue spaces and are defined using Young functions. This limitations of the Orlicz spaces is that we cannot approximate every function with a sequence of smooth, compactly supported functions. We introduce Orlicz-type spaces in short (H-Orlicz spaces), containing Henstock-Kurzweil integrable functions. The theory of H-Orlicz spaces with vector measure is discussed. The weak Henstock-Orlicz spaces are constructed with the locally Henstock-Kurzweil integrable functions, and their inclusion properties are discussed. The countable additivity of the Henstock-Dunford integrable functions is discussed in H-Orlicz spaces. We develop the theory of H-Orlicz space generated by generalized Young function. Modular convergence of H-Orlicz space for the case of vector-valued functions and norm convergence in $\mathcal{H}^{\theta}(\mathcal{X}, \mu_{\infty})$, where $\mathcal{X}$ is any Banach space, are discussed. Relationships of modular convergence and norm convergence of H-Orlicz spaces are discussed.

This monograph is divided into the following chapters.

Chapter 1 We recall few definitions and basic properties of vector-valued functions with respect to scalar measures, and integrals of scalar-valued functions with respect to vector measure. Also, we mention about normed spaces, Banach spaces, and related results. Next, we briefly discuss about Young functions, function spaces, and Orlicz spaces.

Chapter 2 In this chapter, we give a survey of basic results about Orlicz space. Several fundamental properties of Orlicz spaces are discussed. Further, we define a generalized version of Orlicz space called Discrete Orlicz-Morrey spaces. Furthermore, we present the sufficient and necessary conditions for the inclusion properties among these spaces.

Chapter 3 In this chapter, we discuss the Henstock-Kurzweil integral. This is the easiest to learn and best known of those integrals that integrate non-absolutely integrable functions and extend the Lebesgue integral. We provide a fairly detailed account of the Henstock-Kurzweil (HK)-integral and its properties in the one-dimensional case and a brief discussion of the n-dimensional case. In the last, we give brief summary of vector-valued Henstock-Kurzweil integral.

Chapter 4 In this chapter, we investigate some properties and convergence theorem of Kluvánek-Lewis-Henstock μ_∞-integrability for μ_∞-measurable functions that we introduced in [1]. We give a μ_∞-a.e. convergence version of Dominated (resp. bounded) convergence theorem for μ_∞. We introduce Kluvánek-Lewis-Henstock integrable of scalar-valued functions with respect to a set valued measure in a Banach space. Finally, we introduce (KL)-type dominated convergence theorem for the set-valued Kluvánek-Lewis-Henstock integral.

Chapter 5 In this chapter, we investigate the space of the Henstock-Dunford integral function. For any Henstock-Dunford integrable function f, an operator can be constructed such that it is both a continuous linear operator and a weakly compact operator. Furthermore, the adjoint operator is a continuous and weakly compact linear operator for the case of Henstock-Dunford integral. In the end, we introduced Pettis integrability type property for HKP-integrals. We discuss several necessary conditions that $\mathcal{X}$ has HKP-integrability property for weak Baire measure. For a Banach-space-valued function, the necessary and sufficient conditions for the indefinite Henstock-Kurzweil-Pettis (HKP) integral or the Denjoy-Pettis integral to have separable ranges involve properties of the Banach space and the integrability conditions.

Chapter 6 We introduce Orlicz type spaces, namely, Henstock Orlicz spaces (in short H-Orlicz spaces) with Henstock-Kurzweil integrable functions. We prove $C_0^\infty(\mathbb{R}^n)$ is dense subspace of H-Orlicz spaces which are not generally dense in the classical Orlicz spaces.

Chapter 7 In this chapter we extend the theory of Henstock-Orlicz spaces with respect to vector measure. We study the integral representation of operators. Lastly, we study uniform convexity, reflexivity, and the Radon-Nikodym property of the Henstock-Orlicz spaces $\mathcal{H}^\theta(\mu_\infty)$.

Chapter 8 We discuss the structure of H-Orlicz space with locally Henstock integrable functions. The weak Henstock-Orlicz spaces on $\mathbb{R}^n$ and some basic properties of the weak Henstock-Orlicz spaces are studied. We obtain some necessary and sufficient conditions for the inclusion properties of these spaces.

Chapter 9 For a given real Banach space $\mathcal{X}$ and probability space $(\mathcal{B}, \Sigma, \mu_\infty)$, we characterize the countable additivity of Henstock-Dunford integrable function for Henstock integrable function taking values in $\mathcal{X}$ as those weakly measurable function $g : \mathcal{B} \to \mathcal{X}$, for which $\{y^*g \ : y^* \in B^*_{\mathcal{X}}\}$ is relatively weakly compact in some separable Henstock-Orlicz space (in briefly H-Orlicz space) $\mathcal{H}^\theta(\mu)$, where $B^*_{\mathcal{X}}$ is the closed unit ball in $\mathcal{X}^*$. We find relatively weakly compactness of some H-Orlicz space of Henstock-Gel'fand integrable functions.

Chapter 10 In this chapter, we develop the theory of H-Orlicz space generated by generalized Young function. Modular convergence of H-Orlicz space for the case of vector-valued functions and norm convergence in $\mathcal{H}^{\theta}(\mathcal{X}, \mu_{\infty})$, where $\mathcal{X}$ is any Banach space, are discussed. Relationships of modular convergence and norm convergence of H-Orlicz spaces are discussed.

Sehore, Madhya Pradesh, India
Guwahati, Assam, India
March 2025

Hemanta Kalita
Bipan Hazarika

References

1. Boccuto, A., Hazarika, B., Kalita, H.: Kuelbs-Steadman spaces for Banach space-valued measures. Mathematics **8**(6), 1–12 (2020)
2. Thung, T.V.: Some collections of functions Dense in an Orlicz Space. Acta Mathematica Vietnamica **25**(2), 195–208 (2000)

Declarations

Competing Interests The authors have no competing interests to declare that are relevant to the content of this manuscript.

Contents

Notations and Symbols

Sets and Set Functions

- $\mathbb{R}$ denotes set of real numbers.
- If $\mathcal{X}$ is any non-empty set, then $P(\mathcal{X})$ represents its power set.
- A^c is complement of A.
- $A \times B$ is cartesian product of A and B.
- $A \Delta B$ is symmetric difference of A and B.
- $int(A)$ is interior of A.
- Closure of A is $\overline{A}$.
- Let E be a measurable set. We denote χ_E is the characteristic function on E and defined by

$$\chi_E = \begin{cases} 1 \text{ if } x \in E, \\ 0 \text{ if } x \notin E. \end{cases}$$

- We denote $\mu_\infty(E)$ is the Lebesgue measure of E. Variation of μ_∞ is $|\mu_\infty|$, and semivariation of μ_∞ be $||\mu_\infty||$.
- $B(x, r) = \{x \in \mathbb{R} : ||y - x|| < r\}$ is open ball with center x and radius r.
- $B[x, r] = \{x \in \mathbb{R} : ||y - x|| \leq r\}$ is closed ball with center x and radius r.
- If we deal with $\mathbb{R}^n$, then $[\mathbf{a}, \mathbf{b}] \subset \mathbb{R}^n$, and open and closed ball are $B(\mathbf{x}, \mathbf{r})$ and $B[\mathbf{x}, \mathbf{r}]$, respectively.
- $V(F, E)$ denotes variation of a function F on any set E.
- $V_*(F, E)$ denotes weak variation of a function F on any set E.
- BV represents bounded variation.
- AC represents absolutely continuous.
- ACG denotes generalized absolutely continuous.
- ACG_* denotes generalized absolutely continuous for restricted sense.
- $d(a, b)$ is distance between a and b, where $a \in A$ and $b \in B$.
- $H(A, B)$ is Hausdorff metric of A and B.
- WCG denotes weakly compact generated.

Integral and Function Spaces

- We denote θ, Φ, Ψ are Young and N-functions.
- If f is any given function, then $L\int_E f$ means Lebesgue integrable on $E \subset \Sigma$.
- If f is any given function, then $(B)\int_E f$ means Bochner integrable on $E \subset \Sigma$.
- If f is any given function, then $(D)\int_E f$ means Dunford integrable on $E \subset \Sigma$.
- If f is any given function, then $(P)\int_E f$ means Pettis integrable on $E \subset \Sigma$.
- If f is any given function, then $HK\int_E f$ means Henstock-Kurzweil integrable on $E \subset \Sigma$.
- If f is any given function, then $(HD)\int_E f$ means Henstock-Dunford integrable on $E \subset \Sigma$.
- If f is any given function, then $(HKP)\int_E f$ means Henstock-Kurzweil-Pettis integrable on $E \subset \Sigma$.
- $HK(I)$ is the space of Henstock-Kurzweil integrable functions defined on I.
- HK_{loc} denotes the set of locally Henstock-Kurzweil integrable functions.
- $D_{HD}([a,b])$ denotes the space of Henstock-Dunford integrable functions.
- If $\mathcal{X}$ is a Banach space, then $HK([a,b],\mathcal{X})$ is the space of all $\mathcal{X}$-valued Henstock-Kurzweil integrable functions.
- If $\mathcal{X}$ is a Banach space, then $HKP([a,b],\mathcal{X})$ is the space of all $\mathcal{X}$-valued Henstock-Kurzweil-Pettis integrable functions.
- $\mathcal{L}^p_w(\mu_\infty)$ represents weakly p-integrable function spaces.
- Let $\mathcal{X}$ be a Banach space. We denote its norm by $||.||_{\mathcal{X}}$.
- Let Ω be open set in $\mathbb{R}^n$ then $C_c^\infty(\Omega)$ denotes the set of smooth functions with compact support in Ω.
- $\mathbb{R}^n$ is a Banach space under Euclidean norm $||x|| = \sqrt{\sum x_i^2}$.
- $C([0,1]) = \{\text{all continuous functions on } [0,1]\}$ and the uniform norm or sup norm

$$\begin{aligned} ||f||_\infty &= \sup\{|f(x)| : x \in [0,1]\} \\ &= \max\{|f(x)| : x \in [0,1]\}. \end{aligned}$$

- $C_0([0,1]) = \{\text{all continuous functions on } [0,1] \text{ with } f(0) = f(1) = 0\}$.
- $C_c([0,1]) = \{\text{all continuous functions on } [0,1] \text{ vanishing near 0 and 1}\}$.
- $\mathcal{L}^p(\mu)$ is the Lebesgue space with the norm $||f||_p = \left(\int_{\mathcal{X}} |f|^p d\mu_\infty\right)^{\frac{1}{p}}$, $1 \leq p < \infty$.
- $\mathcal{L}^{-\theta}(\mu_\infty)$ represents the Orlicz classes.
- $\mathcal{L}^\theta(\mu_\infty)$ represents the Orlicz spaces.

- M^θ denotes Morse spaces.
- wM^θ denotes weak Morse spaces.
- $l_{\Psi,\theta}(\mathbb{R})$ denotes the Discrete Orlicz Morrey spaces.
- If $\mathcal{B}$ is any abstract set, then $\mathcal{H}^{-\theta}(\mu_\infty)$ represents Henstock-Orlicz classes on $\mathcal{B}$.
- If $\mathcal{B}$ is any abstract set, then $\mathcal{H}^{\theta}(\mu_\infty)$ represents Henstock-Orlicz spaces or H-Orlicz spaces on $\mathcal{B}$.

Chapter 1
Basic Ingredients

This chapter is composed of two parts: Basic Analysis and Functional Analysis. The first part is a review of some of the basic background that is required from the first 2 years of a standard program in mathematics. In the second part of this chapter, we include some intermediate to advanced material that is required later. In most cases, we provide specific references for their proof. One can see [3–5, 11, 15, 21] and their references therein for details of analysis and functional analysis.

1.1 Analysis

Let $\mathcal{X}$ be any non-empty set, let $\emptyset$ be the empty set. Consider $P(\mathcal{X})$ be the power set of $\mathcal{X}$.

Definition 1.1 Let $A, B\ A_n \in P(\mathcal{X}),\ n \in \mathbb{N}$. Then

1. $A^c = \left\{a \in \mathcal{X} : a \notin A\right\}$, the complement of A.
2. $A \setminus B = A \cap B^c$
3. A and B are disjoint if $A \cap B = \emptyset$.
4. (De Morgan's laws) $\left(\bigcup\limits_{k=1}^{\infty} A_k\right)^c = \bigcup\limits_{k=1}^{\infty} A_k^c$, $\left(\bigcap\limits_{k=1}^{\infty} A_k\right)^c = \bigcup\limits_{k=1}^{\infty} A_k^c$.

We define the $\liminf$ and $\limsup$ for sets by: $\lim\limits_{n\to\infty} \inf A_n = \bigcup\limits_{n=1}^{\infty}\bigcap\limits_{k=n}^{\infty} A_k$ and $\lim\limits_{n\to\infty} \sup A_n = \bigcap\limits_{n=1}^{\infty}\bigcup\limits_{k=n}^{\infty} A_n$.

Theorem 1.1 *Let* $\{A_n\} \subset P(\mathcal{X}),\ n \in \mathbb{N}$. *Then* $\liminf$ *and* $\limsup$ *satisfy*

1. $\lim\limits_{n\to\infty} \inf A_n \subset \lim\limits_{n\to\infty} \sup A_n$.

B. Hazarika, H. Kalita, *Theory of Henstock–Orlicz Spaces*, Mathematical Marvels: Texts and Monographs in the Spirit of CR Rao,
https://doi.org/10.1007/978-981-96-9548-5_1

2. $\lim_{n\to\infty} \sup A_n = \{a : a \in A_k \textit{ for infinitely many } k\}$.
3. $\lim_{n\to\infty} \inf A_n = \{a : a \in A_k \textit{ for all but finitely many } k\}$.
4. $\left(\lim_{n\to\infty} \sup A_n\right)^c = \lim_{n\to\infty} \inf A_n^c$.
5. *If* $A_{n+1} \subset A_n$, *then* $\lim_{n\to\infty} \inf A_n = \bigcap_{k=1}^{\infty} A_k$.
6. *If* $A_n \subset A_{n+1}$, *then* $\lim_{n\to\infty} \inf A_n = \bigcup_{k=1}^{\infty} A_k$.

Definition 1.2 Let A, B are nonempty subset of $\mathcal{X}$. The Cartesian product of A, B is

$$A \times B = \{(a, b) : a \in A, \ b \in B\}.$$

Definition 1.3 A function $f : A \to B$ with domain $D(f) \subset A$ and range $R(f) \subset B$ is a subset $f \subset A \times B$ such that for each $x \in A$, there is one and only one $y \in B$ with $(x, y) \in f$.

We called f is one-one (or injective) if for all $x_1 \neq x_2 \in A$ we have $f(x_i) \neq f(x_2) \in B$. We called f is onto or surjective if for each $y \in B$, there is a $x \in A$ with $y = f(x)$.

A metric on a set $\mathcal{X}$ is a function that satisfies the minimal properties we might expect of a distance.

Definition 1.4 A metric d on a set $\mathcal{X}$ is a function $d : \mathcal{X} \times \mathcal{X} \to \mathbb{R}$ such that for all $x, y \in \mathcal{X}$

1. $d(x, y) \geq 0$ and $d(x, y) = 0$ if and only if $x = y$.
2. $d(x, y) = d(y, x)$
3. $d(x, y) \leq d(x, z) + d(z, y)$ for $z \in \mathcal{X}$.

A metric space $(\mathcal{X}, d)$ is a set $\mathcal{X}$ with a metric d defined on $\mathcal{X}$.

Example 1.1 Define $d : \mathbb{R} \times \mathbb{R} \to \mathbb{R}$ defined by $d(x, y) = |x - y|$. Then d is a metric on $\mathbb{R}$.

1.1.1 Topology

We recall about topology. We only consider Hausdorff spaces or spaces with Hausdorff topology .

Definition 1.5 Let $\mathcal{X}$ be a non-empty set and let τ be a set of subsets of $\mathcal{X}$. We say that τ defines a Hausdorff topology on X if

1. $\mathcal{X}, \ \emptyset \in \tau$.

2. If $O_1, O_2, \ldots, O_n$ is a finite collection of sets in τ, then $\bigcap_{i=1}^{n} O_i \in \tau$.
3. If $\bigwedge$ is an index set and for each $\tau \in \bigwedge$, there is a set $O_\gamma \in \tau$, then $\bigcup_{\gamma \in \bigwedge} O_\gamma \in \tau$.
4. If $x, y \in \mathcal{X}$ are two district points, there are two disjoint sets $O_1, O_2 \in \tau$ such that $x \in O_1$ and $y \in O_2$.

A set $N \in \tau$ is called a neighborhood for each point $x \in N$, and the set $\tau_x \subset \tau$ of all neighborhoods for x is called a complete neighborhood basis for x. Thus, any set O, containing x, also contains some neighborhood basis set $N(x) \subset \tau_x$. A set G is said to be closed if G^c is open.

Definition 1.6 Let $A \neq \emptyset$ be a subset of $\mathcal{X}$.

1. The interior of A, denoted by $int(A)$, is the union of all $O \in \tau$ such that $O \subset A$. If $x \in int(A)$, we say that x is an interior point of A.
2. The closure of A, which is denoted by $\overline{A}$, is the set of all $x \in A$ such that for all $N(x) \in \tau_x$, $N(x) \cap A \neq \emptyset$.
3. We say A is dense in $\mathcal{X}$ if $\overline{A} = \mathcal{X}$. If A is countable, we say that $\mathcal{X}$ is separable.

1.1.2 σ-Algebra

Let $\mathcal{A} \subset P(\mathcal{X})$ be a collection of subsets of $\mathcal{X} \neq \emptyset$. We say that $\mathcal{A}$ is an algebra if the following holds:

1. $\mathcal{X}, \emptyset \in \mathcal{A}$.
2. $A, B \in \mathcal{A}$, then $A^c, B^c \in \mathcal{A}$ and $A \cup B \in \mathcal{A}$.
3. $A \cap B \in \mathcal{A}$ and $A \setminus B \in \mathcal{A}$.
4. If n is finite and $\{A_k\} \subset \mathcal{A}$, $1 \leq k \leq n$, then $\bigcup_{k=1}^{n} A_k \in \mathcal{A}$, $\bigcap_{k=1}^{n} A_k = \mathcal{A}$.

Definition 1.7 Let $\mathcal{A} \subset P(\mathcal{X})$ be an algebra. We say that $\mathcal{A}$ is an σ-algebra if $\bigcup_{k=1}^{\infty} A_k \in \mathcal{A}$ for any countable family of sets $\{A_k\} \in \mathcal{A}$.

Also $\bigcap_{k=1}^{\infty} A_k \in \mathcal{A}$ with $\lim_{n \to \infty} \inf A_n \in \mathcal{A}$ and $\lim_{n \to \infty} \sup A_n \in \mathcal{A}$.

Definition 1.8 If Σ is a non-empty class of subsets of $\mathcal{X}$, the smallest σ-algebra $\mathcal{A}$ with $\Sigma \subset \mathcal{A}$ is called the σ-algebra generated by Σ.

If $\mathcal{X}$ is a topological space and Σ is the class of open sets of $\mathcal{X}$, then $\mathcal{A}(\Sigma) = \mathfrak{B}(\mathcal{X})$ is called the Borel σ-algebra of $\mathcal{X}$.

Definition 1.9 Let $\mathcal{X}$ be non-empty set. An outer measure μ_∞^* is a function of $P(\mathcal{X}) \to [0, \infty]$ such that

1. $\mu_\infty^*(\emptyset) = 0$

2. If $B \subset A$, then $\mu_\infty^*(B) \le \mu_\infty^*(A)$
3. If $A \subset \bigcup_{k=1}^{\infty} A_k$, then $\mu_\infty^*(A) \le \sum_{k=1}^{\infty} \mu_\infty^*(A_k)$.

If for each sequence of disjoint sets $\{A_k\} \subset \mathcal{A}$, $\mu_\infty^*\left(\bigcup_{k=1}^{\infty} A_k\right) = \sum_{k=1}^{\infty} \mu_\infty^*(A_k)$, we called $\mu_\infty^* = \mu_\infty$ is a measure.

Definition 1.10 A subset E of $\mathbb{R}$ is Lebesgue measurable if $\mu_\infty^*(T \cap E) + \mu_\infty^*(T \cap E^c) = \mu_\infty^*(T)$ for every $T \subset \mathbb{R}$. In this case, the outer measure of E is called Lebesgue measurable, and we denote $\mu_\infty(E)$.

1.1.3 Functional Analysis

In this section, we include a few basic background results from functional analysis and Banach space theory.

Definition 1.11 A vector space $\mathcal{X}$ over $\mathbb{C}$ is an abelian group under multiplication by elements of $\mathbb{C}$. That is

1. For each $x, y \in \mathcal{X},\ x + y \in \mathcal{X}$.
2. For all $x, y \in \mathcal{X},\ x + y = y + x,\ (x + y) + z = x + (y + z)$.
3. There is a unique element $0 \in \mathcal{X}$ called zero and $x + 0 = 0 = 0 + x\ \forall\, x \in \mathcal{X}$.
4. For all $x \in \mathcal{X}$, there is a unique element $-x \in \mathcal{X}$ and $x + (-x) = (-x) + x = 0$.
5. For all $x, y \in \mathcal{X}$ and $a, b \in \mathbb{C}$, $ax \in \mathcal{X}$; $1x = x,\ (ab)x = a(bx)$ and $a(x + y) = ax + ay$.

A vector space $\mathcal{X}$ is termed as a normed space if for every $x \in \mathcal{X}$ there is a nonnegative real number $||x||$ associated to x, called the norm of x, in such a way that the following holds:

1. $||x + y|| \le ||x|| + ||y||$
2. $||\alpha x|| = |\alpha|||x||$ if $x \in \mathcal{X}$ and α is a scalar.
3. $||x|| > 0$ if $x \ne 0$.

Let $\mathcal{X}$ be a complex linear space. A functional $||.|| : \mathcal{X} \to [0, \infty)$ is called a quasi norm if it satisfies the following condition:

1. It is positive homogeneous, i.e., for all $\alpha\mathbb{C}$, and for all $x \in \mathcal{X}$ $||\alpha x|| = |\alpha|||x||$.
2. It satisfies $||x|| = 0$ if and only if $x = 0$ in $\mathcal{X}$.
3. There is a constant $C \ge 1$ called the modulus of concavity of $||.||$, such that it is subadditive up to this constant. That is for all $x, y \in \mathcal{X}$, $||x + y|| \le C||x|| + ||y||$.

If $C = 1$, the quasi norm is a norm.

Topology on $\mathcal{X}$ is generated by the metric defined by $d(x, y) = ||x - y||$ so that $\{x : ||x - y|| < r\}$ is an open ball about y of radius r. The space $\mathcal{X}$ is complete if every Cauchy sequence in the above norm converges to an element in $\mathcal{X}$. A normed space that is complete in the metric is defined by its norm; this means that every Cauchy sequence is required to converge, which is called a Banach space . We denote $\mathcal{X}^*$ is dual space of $\mathcal{X}$. The topology induced by $d(x, y) = ||x - y||$ is called a norm topology.

A Banach space $\mathcal{X}$ is called reflexive if and only if closed ball is weakly compact. The weak topology on $\mathcal{X}$ is a coarsest (i.e., smallest topology) such that every element of $\mathcal{X}^*$ is continuous. Open (respectively closed) sets in the weak topology will be called weakly open (respectively weakly closed) sets.

Theorem 1.2 *The weak topology is Hausdorff.*

Theorem 1.3 *Let $\mathcal{X}$ be a Banach space and let (x_n) be a sequence in $\mathcal{X}$. Then*

1. *$x_n \rightharpoonup x$ in $\mathcal{X}$ if and only if $f(x_n) \to f(x)$ for all $f \in \mathcal{X}^*$.*
2. *If $x_n \rightharpoonup x$ in $\mathcal{X}$, then $\big(||x_n||\big)$ is bounded and $||x|| \leq \lim\limits_{n\to\infty} \inf ||x_n||$.*

If $\mathcal{X}$ is a finite dimensional, then the norm and weak topologies coincide.

Definition 1.12 The weak* topology on $\mathcal{X}^*$ is the coarsest topology such that the functional $\{J_x : x \in \mathcal{X}\}$ are all continuous, where $J : x \to J_x$ is the canonical imbedding of $\mathcal{X}$ into $\mathcal{X}^{**}$.

Clearly weak* topology is coarsest than the weak topology on $\mathcal{X}^*$. If $\mathcal{X}$ is a reflexive Banach space, then weak and weak* topologies coincide on $\mathcal{X}^*$.

A nonzero Banach space $(\mathcal{X}, ||.||_{\mathcal{X}})$ of measurable functions on $(\mathbb{R}^+, \mu_\infty)$ is called symmetric if the following conditions hold:

(a) If $|f| \leq |g|$ and $g \in \mathcal{X}$, then $f \in \mathcal{X}$ and $||f||_{\mathcal{X}} \leq ||g||_{\mathcal{X}}$.
(b) If f and g are equimeasurable and $g \in \mathcal{X}$, then $f \in \mathcal{X}$ and $||f||_{\mathcal{X}} = ||g||_{\mathcal{X}}$.

A normed space $(\mathcal{X}, ||.||_{\mathcal{X}})$ satisfying condition (a) is called a normed ideal lattice . A partially ordered Banach space $\mathcal{X}$ which is also a vector lattice is a Banach lattice if $\|x\| \leq \|y\|$ for every $x, y \in \mathcal{X}$ such that $|x| \leq |y|$.

A weak order unit of $\mathcal{X}$ is a positive element $e \in \mathcal{X}$ such that, if $x \in \mathcal{X}$ and $x \wedge e = 0$, then $x = 0$.

A Banach lattice $\mathcal{X}$ is said to be order complete (resp. σ-order complete) if every upper order bounded net (resp. sequence) of elements of $\mathcal{X}$ has a supremum in $\mathcal{X}$.

We say that $\mathcal{X}$ is order continuous (resp.) if for each downward-directed set (resp. sequence) $(x_\xi)_{\xi\in\Xi}$ with $\bigwedge\limits_{\xi\in\Xi} x_\xi = 0$, it is $\lim\limits_{\xi\in\Xi} \|x_\xi\| = 0$.

Let $\mathcal{X}$ be a Banach lattice and $\emptyset \neq A \subset B \subset \mathcal{X}$. We say that A is solid in B if for each x, y with $x \in B$, $y \in A$ and $|x| \leq |y|$, it is $x \in A$.

Definition 1.13 [19] A Banach space $(\mathcal{X}, ||.||)$ is uniformly convex if for each $\epsilon > 0$ there exists a $p(\epsilon) \in (0, 1)$ such that $||x|| = ||y|| = 1$, $||x - y|| \geq \epsilon$ implies $||\frac{x+y}{2}|| \leq 1 - p(\epsilon)$.

Let λ be an extended real-valued measure on Σ. A Banach space $\mathcal{X}$ consisting of (classes of equivalence of) λ-measurable functions is called a *Köthe function space* with respect to λ if, for every $g \in \mathcal{X}$ and for each measurable function f with

$|f| \leq |g|$ λ-almost everywhere, it is $f \in \mathcal{X}$ and $\|f\| \leq \|g\|$, and $\chi_A \in \mathcal{X}$ for every $A \in \Sigma$ with $\lambda(A) < +\infty$. Some examples of Banach spaces :

1. $\mathbb{R}^n$ is a Banach space under Euclidean norm $||x|| = \sqrt{\sum x_i^2}$.
2. $C([0, 1]) = \{\text{Continuous functions on } [0, 1]\}$ under the uniform norm or sup norm

$$\begin{aligned} ||f||_\infty &= \sup \{|f(x)| : x \in [0, 1]\} \\ &= \max \{|f(x)| : x \in [0, 1]\}. \end{aligned}$$

Let $\mathcal{X}_1$ be a subspace of the normed space $(\mathcal{X}, ||.||)$. Then $(\mathcal{X}_1, ||.||)$ is again normed space. The following are some subspaces in $C([0, 1])$:

1. $C_0([0, 1]) = \{\text{All continuous functions with } f(0) = f(1) = 0\}$.
2. $C_c([0, 1]) = \{\text{All continuous functions on } (0, 1) \text{ vanishing near 0 and 1}\}$.

Let $(\mathcal{X}, \Sigma, \mu_\infty)$ be a measure space and $0 < p < \infty$. A measurable function f is called p-integrable function if $|f|^p$ is integrable.

Theorem 1.4 (Hölder Inequality) *Let f and g be measurable. For $1 < p < \infty$, $\int_{\mathcal{X}} |fg| d\mu_\infty \leq \left(\int_{\mathcal{X}} |f|^p d\mu_\infty \right)^{\frac{1}{p}} \left(\int_{\mathcal{X}} |g|^q d\mu_\infty \right)^{\frac{1}{q}}$ where q is conjugate to p.*

Moreover, when the right-hand side of this inequality is finite, equality sign holds if and only if either

1. f or g vanishes a.e. or
2. There are some $\alpha, \beta \geq 0$, $\alpha\beta \neq 0$ such that $\alpha|f|^p = \beta|g|^q$ a.e.

Theorem 1.5 (Minkowski's Inequality) *Let f and g be measurable. For $1 \leq p < \infty$ following inequality hold: $\left(\int_{\mathcal{X}} |f + g|^p d\mu_\infty \right)^{\frac{1}{p}} \leq \left(\int_{\mathcal{X}} |f|^p d\mu_\infty \right)^{\frac{1}{p}} + \left(\int_{\mathcal{X}} |g|^p d\mu_\infty \right)^{\frac{1}{p}}$.*

Let $\mathcal{L}^p(\mu_\infty) = \{p - \text{integrable functions on } (\mathcal{X}, \Sigma, \mu_\infty)\}$. This space is called Lebesgue space. The Lebesgue space $\mathcal{L}^p(\mu)$ is a Banach space with the norm defined below: $||f||_p = \left(\int_{\mathcal{X}} |f|^p d\mu_\infty \right)^{\frac{1}{p}}$, $1 \leq p < \infty$.

1.1.4 Integration Theory

Integration theory of functions with values in a Banach space has long been a fruitful area of study. Bochner, Gelfand, Pettis, Birkhoff, Phillips, etc. have major contributions in this area of interest. Edgar, Huff, and Talagrand discussed Pettis integrability with scalar measurability. For Pettis integrability of weak measurable functions with values by several different techniques, see [22–24]. We recall a few definitions and basic properties of vector valued functions with respect to scalar measures and integrals of scalar-valued functions with respect to vector measure. The measurability of functions are strong and weak measurability. Let $(\mathcal{X}, \Sigma, \mu_\infty)$ a measure space.

1. μ_∞ is finite measure if $\mu_\infty(X) < \infty$.
2. μ_∞ is concentrated on a set $A \in \Sigma$ if $A = U^c$ and U is the largest open set with the property $\mu_\infty(U) = 0$.
3. μ_∞ is regular measure if given $A \in \Sigma$, for each $\epsilon > 0$, there is an open set O and a closed set K such that $K \subset A \subset O$ and $\mu_\infty(O \setminus K) < \epsilon$.
4. We say μ_∞ is a σ-finite measure if there is a sequence $\{A_k\} \subset \Sigma$ with $X = \bigcup_{k=1}^{\infty} A_k$ and $\mu_\infty(A_k) < \infty$.

Let $(\mathcal{X}, \Sigma, \mu_\infty)$ be a measure space. Then

1. Let f be a function on X, $f : X \to K$ where $K = \mathbb{R}$ or $\mathbb{C}$.
2. f is measurable if $f^{-1}(B) \in \Sigma$ for every $B \in \mathfrak{B}(K)$, the Borel algebra on K.
3. Two functions f and g are equal almost everywhere $f = g$ μ_∞-a.e. if they have same domain and $\mu_\infty\{x : f(x) \neq g(x)\} = 0$. In general, a property is said to hold μ_∞-a.e. on X if the set of points where this property fails has μ_∞-measure zero.

Definition 1.14 A function $f : \Omega \to \mathcal{X}$ is called simple if there exist $x_1, x_2, .., x_n$ and $E_1, E_2, .., E_n \in \Sigma$ such that $f = \sum_{i=1}^{n} x_i \chi_{E_i}$, where χ_{E_i} is a characteristic function on E_i.

A function $f : \Omega \to \mathcal{X}$ is called μ_∞-measurable if there exists a sequence of simple function (f_n) with $\lim_n ||f_n - f|| = 0$ μ_∞ a.e. A function $f : \Omega \to \mathcal{X}$ is called weakly μ_∞-measurable if for each $x^* \in \mathcal{X}^*$, $x^* f$ is μ_∞-measurable. If $f : \Omega \to \mathcal{X}^*$ is $\mathcal{X}$-measurable ($\mathcal{X}$ is identified with its image under natural embedding of $\mathcal{X}$ into $\mathcal{X}^{**}$) is called weak*-measurable . For every nonnegative measurable function $f : \mathbb{R}^+ \to \mathbb{R}^+$, we define the function

$$\eta_f(t) := \mu_\infty(\{f > t\}), t \in \mathbb{R}^+, \tag{1.1}$$

where $\{f > t\} := \{x \in \mathbb{R}^+ : f(x) > t\}$, $t \in \mathbb{R}^+$ are the upper Lebesgue sets of f with the Lebesgue measure μ_∞.

The function $\eta_f : \mathbb{R}^+ \to [0, \infty]$, defined by (1.1) is called the (upper) distribution function of f. Clearly η_f is smooth (infinitely differentiable) and η_f has compact support (is identically zero outside some bounded interval).

The nonnegative functions f and g are called equimeasurable if $\eta_f = \eta_g$, i.e., $\mu_\infty(\{f > t\}) = \mu_\infty(\{g > t\}),\ t \in \mathbb{R}^+$.

Let $f : \mathbb{R}^+ \to \mathbb{R}$ be a real measurable function on $\mathbb{R}^+$ and let $\eta_{|f|}(t) := \mu_\infty(\{|f| > t\}) \in [0, +\infty], t \in \mathbb{R}^+$ be the distribution function of the absolute value $|f|$. It is possible to have $\eta_{|f|}(t) = +\infty$. In the sequel, we assume (unless otherwise stated) that $\eta_{|f|}(t) \neq +\infty$. Since the function $\eta_{|f|}$ is decreasing and right-continuous, it has a unique generalized inverse, which is also decreasing and right-continuous. This inverse function $\eta_{|f|}^{-1}$ will be denoted by

$$\eta_{|f|}^{-1}(x) = \inf\{t \geq 0 : \eta_{|f|}(t) \leq x\}.$$

Since $\eta_{|f|}(t) \neq +\infty$, there exists $t_0 \geq 0$ such that the value $\eta_{|f|}(t_0) = \mu_\infty(\{|f| > t_0\}) < +\infty$. This means that $\lim_{t\to\infty} \eta_{|f|}(t) = 0$ and hence $\eta_{|f|}^{-1}(x) < \infty$ for all $x > 0$.

Definition 1.15 ([10, Definition 1]) A μ_∞-measurable function $f : \Omega \to \mathcal{X}$ is called Bochner integrable if there exists a sequence of simple-functions (f_n) such that

$$\lim_n \int_\Omega ||f_n - f|| d\mu_\infty = 0.$$

In this case $\int_E f d\mu_\infty$ is defined for each $E \in \Sigma$ by

$$\int_E f d\mu_\infty = \lim_n \int_E f_n d\mu_\infty,$$

where $\int_E f_n d\mu_\infty$ is defined in the obvious way.

A concise characterization of Bochner integrable functions is "A μ_∞-measurable function $f : \Omega \to \mathcal{X}$ is Bochner integrable if and only if $\int_\Omega ||f|| d\mu_\infty < \infty$" (see [10, Theorem 2]). The theory of integration similar to the Bochner integrals for weakly measurable function is not possible.

Definition 1.16 A function $f : [0, 1] \to \mathcal{X}$ is said to be scalarly measurable (resp scalarly integrable) if for every $x^* \in \mathcal{X}^*$ the function $x^* f$ is Lebesgue measurable (resp integrable). A scalarly integrable function $f : [0, 1] \to \mathcal{X}$ is said to be Pettis integrable if for each $A \in \Sigma$ there exists a vector $w_A \in \mathcal{X}$ such that

$$< x^*, w_A >= \int_A x^* f d\mu_\infty \text{ for every } x^* \in \mathcal{X}^*.$$

We call w_A the Pettis integral of f over A and we write $w_A = (P) \int_A f d\mu_\infty$.

Plebanek [20] has shown that the Pettis integral property of a Banach space $\mathcal{X}$, together with the requirement of separability of $\mathcal{X}$-valued Pettis integral, is equivalent to the fact that every weak Baire measure on $\mathcal{X}$ is, in a certain weak sense, concentrated on a separable subspace. Recall a Banach space $\mathcal{X}$ is said to have Pettis integral property (PIP) if for every $\mathcal{X}$-valued weakly bounded function f defined on an arbitrary finite measure space is Pettis integrable. We say that $\mathcal{X}$ has Mazur's property if each weak*-continuous functional on $\mathcal{X}^*$ is in $\mathcal{X}$.

Definition 1.17 A scalarly measurable function $f : [0, 1] \to \mathcal{X}$ is determined by a space $Y \subseteq \mathcal{X}$ if for every $x^* \in Y^{\perp}$ the equality $x^* f = 0$ hold μ_∞-a.e.

Dunford proposed a Lemma as follows:

Lemma 1.1 ([10, Lemma 1]) *Suppose f is a weakly μ_∞-measurable function on Ω and $x^* f \in \mathcal{L}^1(\Omega)$ for each $x^* \in \mathcal{X}^*$. Then for each $E \in \Sigma$ there exists $x_E^{**} \in \mathcal{X}^{**}$ satisfying $x_E^{**}(x^*) = \int_E x^*(f) d\mu_\infty$, for all $x^* \in \mathcal{X}^*$.*

With the help of Lemma 1.1, Dunford and Pettis integral is defined as follows:

Definition 1.18 If f is weakly μ_∞-measurable $\mathcal{X}$-valued function on Ω such that $x^* f \in \mathcal{L}^1(\mu_\infty)$ for all $x^* \in \mathcal{X}^*$, then f is Dunford integrable and the Dunford integrable of f over $E \in \Sigma$ is defined by the element x_E^{**} such that for all $x^* \in \mathcal{X}^*$, $x_E^{**}(x^*) = \int_E x^* f d\mu_\infty$, and $x_E^{**} = (D) \int_E f d\mu_\infty$.

Next we discuss about integrals through vector measure. Measure theory was developed in successive stages during the late nineteenth and early twentieth centuries by Emile Borel, Henri Lebesgue, Johann Radon, and Maurice Frechet, among others. In integration theory, specifying a measure allows one to define integrals on spaces more general and has a richer theory than its predecessor. Vector measure is a function defined on a family of sets and taking vector values satisfying certain properties (we mention in next subsection).

Definition 1.19 A function μ_∞ from a field F of subsets of a set Ω to a Banach space $\mathcal{X}$ is called a finitely additive vector measure or simply a vector measure if whenever A_1, A_2 are disjoint members of F, then $\mu_\infty(A_i \cup A_2) = \mu_\infty(A_1) + \mu_\infty(A_2)$. In addition $\mu_\infty(\bigcup_{i=1}^{\infty} A_i) = \sum_{i=1}^{\infty} \mu_\infty(A_i)$ in the norm topology of $\mathcal{X}$ for all sequence (A_i) of pairwise disjoint members of F such that $\bigcup_{i=1}^{\infty} A_i \in F$, then μ_∞ is termed as a countable additive vector measure or simply μ_∞ is countable additive.

We state an example of finitely additive and countably additive vector measures.

Example 1.2 Let Σ be a σ-field of subsets of a set Ω, λ be a nonnegative countably additive measure on Σ. Consider $1 \leq p \leq \infty$. Define $(\mu_\infty)_p : \Sigma \to \mathcal{L}^p(\lambda)$ by $(\mu_\infty)_p(A) = \chi_A$ for each $A \in \Sigma$, where χ_A is characteristic function on A.

Basic theory and isomorphic theory of Banach spaces have vector measure-theoretic origins. J. A. Clarkson (1906–1970) in 1936 introduced the notion of uniform convexity to prove continuous functions on a Euclidean space with values in a uniformly convex Banach space are the integrals of their derivatives. At the same time, Clarkson used vector measure-theoretic ideas to prove that much familiar Banach spaces don't admit equivalent uniformly convex norms. Nelson James Dunford (December 12, 1906–September 7, 1986) and Anthony Perry Morse (1911–1984), in 1936, introduced the notion of a boundedly complete basis are the integrals of their derivatives. Shortly thereafter Dunford was able to recognize the Dunford-Morse theorem and the Clarkson theorem as genuine Radon-Nikodym theorem for the Bochner integral. This was the first Radon-Nikodym theorem for vector measures on abstract measure spaces. Billy James Pettis (1913–April 14, 1979) in 1938 made his contribution to the Orlicz-Pettis theorem to prove that weakly countably additive vector measures are norm countably additive. There was like a full stop in the area of vector measure during World War II. After World War II the work in the vector measure theory became a little more formal generalization of the scalar theory. Representation theory for operators on function spaces became the vogue. During the 1950s and early 1960s, the theory of vector measures languished in sterility. There were one or two bright spots in the mid-1950s. In the mid-1950s, Alexander Grothendieck (1928–2014) used the vector measure theory of the late 1930s and early 1940s to launch a monumental study of linear operators. R.G. Bartle (1927–2003), N. Dunford, and J.T. Schwartz (1930–2009) studied operators on spaces of continuous functions and proved the first important theorems in the theory of vector measures in some 15 years except Igor Kluvánek (1931–1993) and few others followed, but unfortunately, no one followed their lead. In the early 1960s, largely through the pioneering work of A. Pelezyrimski and J. Lindenstrauss, Banach's space theory came back to life and today has reemerged as a deep and vigorous area of mathematical inquiry. In the mid-1960s, N. Dinueleanu gave an intensive study of many of the theorems of vector measure theory that had been proven between 1950 and 1965. At the beginning of the second part of twentieth century, Bartle-Dunford-Schwartz [2] introduced the theory of scalar functions for a countable additive vector measure $\mu_\infty : \Sigma \to \mathcal{X}$ with values in a Banach space $\mathcal{X}$, and studied by Kluvánek-Knowles [16] and Lewis [17]. Fernandez, Fernandez-Naranjo, Okada, Rodriguez, L. Piazza, Stefansson, and many more studied and analyzed the properties of the space of integrable functions $\mathcal{L}^p(\mu_\infty)$, see [6–8, 12, 13]. Antonio Fernández, Fernando Mayoral, Francisco Naranjo, Carman Saéz, and E.A. Sanchez Perez in [14] studied several properties of the Banach lattices $\mathcal{L}^p(\mu_\infty)$ and $\mathcal{L}^p_w(\mu_\infty)$ of p-integrable scalar functions and weakly p-integrable scalar functions for a countably additive vector measure μ_∞. The relation between these two spaces plays a fundamental role in our analysis. Ricardo del Campo, Antonio Fernandez, Fernando Mayoral, and Francisco Naranjo analyzed the localization of uniformly integrable subsets of scalar integrable functions for μ_∞ in a suitable Orlicz space associated with μ_∞. As tools they considered the properties of Orlicz spaces associated with intermediate functions of a given pair of Young functions. These properties allowed them to obtain compactness properties of the inclusion operators, see [9]. More properties of vector measures are discussed later in the respective chapters.

References

1. Alexiewicz, A.: Linear functionals on Denjoy-integrable functions. Collectanea Mathematica **1**, 289–293 (1948)
2. Bartle, R.G., Dunford, N., Schwartz, J.: Weak compactness and vector measures. Can. J. Math. **7**, 289–305 (1955)
3. Bartle, R.G., Sherbert, D.R.: Introduction to Real Analysis 3rd edn. Wiley-India (2010)
4. Brézis, H.: Functional Analysis, Sobolev Spaces and Partial Differential Equations. Springer, New York-Dordrecht-Heidelberg-London (2011)
5. Bruckner, A.M., Bruckner, J.B., Thomson, B.S.: Real Analysis, 2nd edn. Classical Real Analysis.com (2008).
6. Curbera, G.P., Ricker, W.J.: Vector Measures, Integration and Applications, Trends in Mathematics, Positivity, pp. 127–160 (2007)
7. Curbera, G.P.: When L^1 of a vector measure is an AL-space. Pacific. J. Math. **162**(2), 287–303 (1994)
8. Curbera, G.P.: Banach space properties of L^1 of a vector measure. Proc. Am. Math. Soc. **123**(12), 3797–3797 (1995)
9. del Campo, R., Fernández, A., Mayoral, F., Naranjo, F.: The de la Vallée-Poussin theorem and Orlicz spaces associated to vector measure. J. Math. Anal. Appl. **470**, 279–291 (2019)
10. Diestel, J., Uhl, J.J.: Vector measures. American Mathematical Society, Providence (1977)
11. Dunford, N., Schwartz, J.T.: Linear Operators Part I. Wiley-Interscience, New York (1988)
12. Federson, M.: Some Peculiarities of the HK-integrals of Banach space-valued functions. Real Anal. Exchange **29**(1), 439–460 (2003/2004)
13. Fernandez, A., Mayoral, F., Naranjo, F., Paual, P.J.: Weakly sequentially complete Freáchet spaces of integrable functions. Archiv der Mathematik **71**(3), 223–228 (1998)
14. Fernández, A., Mayoral, F., Naranjo, F., Sáez, C., Sánchez-Pérez, E.A.: Spaces of p-integrable functions with respect to a vector measure. Positivity **10**, 1–6 (2006)
15. Gill, T.L., Zachary, W.W.: Functional Analysis and Feynman Operator Calculus. Springer, Cham (2016)
16. Kluvánek, I., Knowles, G.: Vector Measures and Control Systems. Mathematics Studies. North Holland Publishing Company, Amsterdam (1975)
17. Lewis, D.R.: Integration with respect to vector measures. Pac. J. Math. **33**(1), 157–165 (1970)
18. Lindenstrauss, J., Tzafriri, L.: Classical Banach Spaces II, Function Spaces. A Series of Modern Surveys in Mathematics. Springer, Berlin-Heidelberg-New York (1979)
19. Luxemburg, W.A.J.: Banach Function Spaces, Thesis, Delft (1955)
20. Plebanek, G.: On Pettis integrals with separable range. Colloquium Mathematicae **64**(1), 71–78 (1993)
21. Rubshtein, B.Z.A., Grabarnik, G.Y., Muratov, M.A., Pashkova, Y.S.: Foundations of Symmetric Spaces of Measurable Functions. Developments in Mathematics. Springer, Berlin (2016)
22. Skaff, M.S.: Vector valued Orlicz spaces generalized N-function, I. Pac. J. Math. **28**(1), 193–206 (1969)
23. Skaff, M.S.: Vector valued Orlicz spaces, II. Pac. J. Math. **28**(2), 1–18 (1969)
24. Talagrand, M.: Pettis integral and measure theory. Mem. Am. Math. Soc. **51**, 2–25 (1984)

Chapter 2
Orlicz Spaces

In this chapter, we give a survey of basic results about Orlicz space. Several fundamental properties of Orlicz spaces are discussed. Further we define a generalized version of Orlicz space called Discrete Orlicz-Morrey spaces. Furthermore, we present the sufficient and necessary conditions for the inclusion properties among these spaces.

2.1 Introduction

The history of function spaces is extensive. They are crucial to both classical and contemporary mathematics. Spaces whose elements are continuous, or differentiable, or p-integrable functions are of interest for their own sake. They are also useful tools for the study of ordinary and partial differential equations. Ever since Lebesgue's theory of integration has taken center stage in concrete problems of analysis, the need for a more inclusive class of spaces than the $\mathcal{L}^p$ families naturally arose. In retrospect, this could be seen as early as 1915 with the publications of de la Vallée Poussin, although it is the Banach space research of the 1920s that formally gave birth to what is later called the Orlicz spaces, first proposed by Z.W. Birnbaum and W. Orlicz and immediately followed and developed by Orlicz himself. However, the necessary tools centering around the convex function theory were already well established in a study of the Fourier series by W.H. Young, Jensen, and others early in the twentieth century. A systematic study of Orlicz spaces has taken place from the 1930s, and after the war years, the study and applications have been vigorous in Poland, USSR, and Japan, the latter under the leadership of H. Nakano with the name "modular spaces."The work of Zaanen, and especially his book on linear analysis in 1953, has enabled Western countries to develop the theory and applications in many directions. Particularly, the monograph by M.A. Krasnoselskii and Ya. B. Rutickii, available by 1961 in English, devoted entirely to Orlicz spaces,

B. Hazarika, H. Kalita, *Theory of Henstock–Orlicz Spaces*, Mathematical Marvels: Texts and Monographs in the Spirit of CR Rao,
https://doi.org/10.1007/978-981-96-9548-5_2

has become a catalyst in having the theory used effectively in differential and integral equations, probability, statistics, as well as in harmonic analysis. In 1986, Julian Musielak, in collaboration with C. Wu, T. Wang, and Y. Wang published a book, *Theory of Geometry of Orlicz Spaces* (in Chinese), which collected the main results on the geometry of Orlicz spaces as well as some applications obtained by that time. It has also been instrumental in suggesting some new ideas in the abstract theory of topological vector spaces. At the same time, we can say that after the appearance of Orlicz spaces, the Theory of $\mathcal{L}^p$-spaces has progressed in several directions. The special structure of Orlicz spaces raises several new questions that are hidden in the classical theory because of the special properties of $\mathcal{L}^p$-spaces. Orlicz spaces can be the first step to consider more abstract generalizations to the classical theory of Banach spaces. The main interest of the application of Orlicz spaces is Potential Theory, Interpolation, and Differential Equations, among others. The Lebesgue integration theory was developed in the year 1904 by Henri Lebesgue. This theory and the function x^p in the definition of $\mathcal{L}^p$ space were inspired by Z.W. Birnbaum and W. Orlicz to propose a generalized space of $\mathcal{L}^p$; later on it was known as Orlicz space. This space was later developed by Orlicz himself. The fundamental properties of Orlicz space with Lebesgue measure are found in [10]. In [10], Rao and Ren discussed the theory of Orlicz space, which is a more generalized version of $\mathcal{L}^p$ space with the help of Young functions and the underlying measure. The basic ideas of the proofs of the theorems of Orlicz space were analogs of the basic results on $\mathcal{L}^p$-space. In [28], Thung presented a translation-invariant subspace $\mathcal{L}^1(\mathbb{R}^n) \cap \mathcal{L}^\phi(\mathbb{R}^n)$ to be dense in Orlicz space $\mathcal{L}^\phi(\mathbb{R}^n)$ by overcoming difficulties as $C_0^\infty(\mathbb{R}^n)$ is dense in $\mathcal{L}^p(\mathbb{R}^n)$ but not generally dense in $\mathcal{L}^\phi(\mathbb{R}^n)$. On the other hand, Donaldson and Trudinger [4] proved that $C_0^\infty(\mathbb{R}^n)$ is dense in Orlicz-Sobolev space. Also for details on Orlicz space and its applications, one can see [10, 19, 22, 25] and [5–10, 13, 15–18, 20, 21, 23, 24, 26, 27]. In the recent period, many researchers have engaged themself to extend the theory of Orlicz spaces to several generalized manners. One can see [4] for Orlicz-Sobolev spaces and [3, 18] for Orlicz-Morrey spaces for details.

The structure of the chapter is as follows: In Sect. 2.2, we discuss the structure of Orlicz class and several results of Orlicz class. In Sect. 2.3, we discuss some of the very important properties of Orlicz spaces, namely completeness, separability, and reflexivity. We also discuss several properties of weak Orlicz spaces. In Sect. 2.4, we discuss several inclusion properties of Orlicz and weak Orlicz spaces. In the last section, we give an introduction of generalized Orlicz spaces, namely Orlicz-Morrey spaces. Further, another version of Discrete Orlicz-Morrey spaces has been defined. Furthermore, we present the sufficient and necessary conditions for the inclusion properties among the Discrete Orlicz-Morrey spaces and weak Discrete Orlicz-Morrey spaces.

2.2 Construction of Orlicz Classes

In this section of the chapter, we discuss about Orlicz class. The relationship of Orlicz classes with classical Lebesgue spaces is discussed. Before coming to the construct of Orlicz class, we recall some definitions and properties of Young functions.

Definition 2.1 ([10]) A function $\theta : \mathbb{R} \to \mathbb{R}^+$ is said to be Young function which satisfies the conditions $\theta(-x) = \theta(x),\ \theta(0) = 0, \theta(x) \to \infty$ as $x \to \infty$.

Definition 2.2 ([10]) A nice Young function $\theta : \mathbb{R} \to \mathbb{R}^+$ is called an N-function and is a continuous Young function such that $\theta(x) = 0$ if and only if $x = 0$ and

$$\lim_{x\to\infty} \frac{\theta(x)}{x} = +\infty \text{ and } \lim_{x\to 0} \frac{\theta(x)}{x} = 0 \ \ \textit{while}\ \theta(\mathbb{R}) \subset \mathbb{R}^+.$$

For example, $\theta_p(x) = x^p;\ p > 1$.

Let us fix a positive finite measure μ_∞ and let θ be an N-function.

Definition 2.3 ([10])

(a) An N-function θ is said to satisfy Δ' condition if there is a $c > 0$ such that

$$\theta(xy) \leq c\theta(x)\theta(y) \text{ for large values of } x \text{ and } y.$$

(b) An N-function θ is said to satisfy Δ_2 condition if there is a $c > 0$ so that

$$\theta(2x) \leq c\theta(x) \text{ for large values of } x.$$

We assume $\mathfrak{F}$ be the class of Young function $\theta : [0, \infty) \to [0, \infty)$ be an increasing, bijective, continuous, and concave satisfying $\theta(0) = 0;\ \lim_{t\to 0} \theta(t) = 0$ and $\lim_{t\to\infty} \theta(t) = \infty$. We denote $\theta_1 \Delta \theta_2$ for $\theta_1, \theta_2 \in \mathfrak{F}$ if there is a constant $c > 0$ such that $\theta_1(t) \leq \theta_2(ct)$ for all $t \geq 0$.

Definition 2.4 ([10, 24]) Let $\chi : [0, \infty] \to [0, \infty]$ be a nondecreasing left continuous function such that $\chi(0) = 0$. The inverse function ξ of χ is defined by $\xi(0) = 0$ and $\xi(v) = \sup\{u : \chi(u) < v\}$. If $\lim_{u\to\infty} \chi(u) = A < \infty$, then $\xi(v) = \infty$ for all $v > A$. The functions defined below

$$\theta(u) = \int_0^u \chi(t)dt \ \textit{and}\ \phi(v) = \int_0^v \xi(t)dt$$

are called conjugate Young function and they satisfy the Young's inequality

$$uv \leq \theta(u) + \phi(v),\ u, v \geq 0.$$

Some properties of Young (N-functions) are listed below. Details of proofs can be seen in [10, 22].

Theorem 2.1 *Let (θ, ϕ) be a complementary pair of N-functions. Then θ, ϕ are strictly increasing so that their inverses θ^{-1}, ϕ^{-1} are uniquely defined and*

1. $\theta(a) + \theta(b) \leq \theta(a+b)$; $\theta^{-1}(a+b) \leq \theta^{-1}(a) + \theta^{-1}(b)$ *for* $a, b \in \mathbb{R}^+$.
2. $a < \theta^{-1}(a)\phi^{-1}(a) \leq 2a$, $a > 0$.

Theorem 2.2 *If (θ_i, ϕ_i), $i = 1, 2$ are two complementary pairs of Young functions and $\theta_1(x) \leq \theta_2(x)$, $x \geq x_0 \geq 0$, then $\theta_2(y) \leq \phi_1(y)$ for all $y \geq y_0 \geq 0$ where $y_0 = \phi(x_0)$.*

Theorem 2.3 *Let θ be an N-function with ϕ as its complementary function. If Φ, Ψ are the (left derivatives) of θ and ϕ, then the following are equivalent:*

1. $\theta \in \Delta_2$.
2. $\exists\, 1 < \alpha < \infty$ *and* $x_0 \geq 0$ *such that* $\frac{x\Phi(x)}{\theta(x)} < \infty$, $x \geq x_0$.
3. $\exists\, 1 < \beta < \infty$ *and* $y_0 \geq 0$ *such that* $\frac{y\phi(y)}{\phi(y)} > \beta$, $y \geq y_0$.

Since a Young function is an extended real Borel function, so if $f : \mathcal{B} \to \mathbb{R}$ is a measurable function, then $\theta(f) : \mathcal{B} \to \mathbb{R}^+$ is a measurable.

Definition 2.5 ([10, 22]) A set of all measurable function $f : \mathcal{B} \to \mathbb{R}$ for Σ is called Orlicz class if $\int_{\mathcal{B}} \theta(|f|) d\mu_\infty < \infty$.

We deonte an Orlicz class by $\overline{\mathcal{L}}^\theta(\mu_\infty)$. That is

$$\overline{\mathcal{L}}^\theta(\mu_\infty) = \left\{ f : \mathcal{B} \to \mathbb{R} \text{ measurable for } \Sigma, \text{ s.t. } \int_{\mathcal{B}} \theta(f) d\mu_\infty < \infty \right\}.$$

Even though only real functions are considered in Orlicz class $\overline{\mathcal{L}}^\theta(\mu_\infty)$, the basic structure theory extends to the complex valued case also by considering the real and imaginary parts separately. In the studies of Orlicz spaces, mostly we assume that measures μ_∞ have the finite subset property, i.e., $E \in \Sigma$, $\mu_\infty(E) > 0$ implies there exists $F \in \Sigma$, $F \subset E$ and $0 < \mu_\infty(F) < \infty$.

Theorem 2.4 ([10, 22]) *The space $\overline{\mathcal{L}}^\theta(\mu_\infty)$ is absolutely convex. That is for $f, g \in \overline{\mathcal{L}}^\theta(\mu_\infty)$ and α, β are scalars such that $|\alpha| + |\beta| \leq 1$ implies $\alpha f + \beta g \in \overline{\mathcal{L}}^\theta(\mu_\infty)$. Also $h \in \overline{\mathcal{L}}^\theta(\mu_\infty)$, $|f| \leq |h|$, f measurable implies $f \in \overline{\mathcal{L}}^\theta(\mu_\infty)$.*

Linearity of any space is an important property. The growth property Δ_2 of a Young function plays an important role to find linearity of Orlicz space.

Theorem 2.5 ([22]) *$\overline{\mathcal{L}}^\theta(\mu_\infty)$ is a linear space if $\theta \in \Delta_2$ globally when $\mu_\infty(\mathcal{B}) = \infty$, and locally if $\mu_\infty(\mathcal{B}) < \infty$. Conversely Δ_2-condition is necessary if μ_∞ is diffused on a set of positive measure.*

Definition 2.6 ([10, 22]) Orlicz class $\overline{\mathcal{L}}^{\theta}(\mu_{\infty})$ is called a circled and solid space if for $f \in \overline{\mathcal{L}}^{\theta}(\mu_{\infty})$, the following condition holds:

1. $h, g \in \overline{\mathcal{L}}^{\theta}(\mu_{\infty})$ where h is any scalar of absolute value.
2. $|g| \leq |f|$, g measurable implies $g \in \overline{\mathcal{L}}^{\theta}(\mu_{\infty})$.

Theorem 2.6 ([10, 22]) *The set $\overline{\mathcal{L}}^{\theta}(\mu_{\infty})$ is a circled solid class of scalar functions, and it is linear if and only if it is closed under multiplication by positive numbers.*

Next, we look into the relationship of classical Lebesgue spaces $\mathcal{L}^1(\mu_{\infty})$ and $\mathcal{L}^{\infty}(\mu_{\infty})$ with Orlicz class.

Theorem 2.7 ([10, 22])

1. $\mathcal{L}^1(\mu_{\infty}) = \bigcup\left\{\overline{\mathcal{L}}^{\theta}(\mu_{\infty}) : \theta \textit{ ranges of all continuous Young function}\right\}$.
2. $\mathcal{L}^{\infty}(\mu_{\infty}) = \bigcap\left\{\overline{\mathcal{L}}^{\theta}(\mu_{\infty}) : \theta \textit{ ranges of all continuous Young function}\right\}$.

Moreover, if (θ, ϕ) is a complementary pair of continuous Young functions, then $\mathcal{L}^1(\mu_{\infty}) = \overline{\mathcal{L}}^{\phi}(\mu_{\infty}).\overline{\mathcal{L}}^{\theta}(\mu_{\infty})$, *where* $\overline{\mathcal{L}}^{\theta}(\mu_{\infty})\overline{\mathcal{L}}^{\phi}(\mu_{\infty}) = \left\{fg : f \in \overline{\mathcal{L}}^{\theta}(\mu_{\infty}), g \in \overline{\mathcal{L}}^{\phi}(\mu_{\infty})\right\}$.

2.3 Orlicz Spaces and Their Properties

Let us recall the formal definition of Orlicz space as below.

Definition 2.7 ([22, Def 3.1.5 page 49]) Suppose an arbitrary measure space $(\mathcal{B}, \Sigma, \mu_{\infty})$, and a Young function θ, for some $\alpha > 0$ the function space for some $\alpha > 0$

$$\mathcal{L}^{\theta}(\mu_{\infty}) = \left\{f : \mathcal{B} \to \mathbb{R},\ \textit{measurable} : \int_{\mathcal{B}} \theta(\alpha f)d\mu_{\infty} < \infty\right\}$$

is called an Orlicz space.

Theorem 2.8 ([10, 22]) *The set $\mathcal{L}^{\alpha}(\mu_{\infty})$ is a vector space. Moreover, for each $f \in \mathcal{L}^{\theta}(\mu_{\infty})$, there is an $\alpha > 0$ such that $\alpha f \in B_{\theta} = \left\{g \in \overline{\mathcal{L}}^{\theta}(\mu_{\infty}) : \int_{\mathcal{B}}\theta(g)d\mu_{\infty} \leq 1\right\}$ and B_{θ} is a circled solid subset of $\overline{\mathcal{L}}^{\theta}(\mu_{\infty})$.*

There are two norms on Orlicz space. They are Luxemburg norm defined by Luxemburg in 1955 and Orlicz norm defined by Nakano in 1950 (see [14]). The Luxemburg norm and the Orlicz norm of the Orlicz space are

$$||.||_\theta = \inf\left\{\alpha > 0 : \int_{\mathcal{B}} \theta\left(\frac{|f|}{\alpha}\right) \leq 1\right\} \tag{2.1}$$

$$||.||_{(\theta)} = \sup\left\{\int_{\mathcal{B}} |fg| d\mu_\infty : \int_{\mathcal{B}} \phi(|g|) d\mu_\infty \leq 1\right\} \tag{2.2}$$

These two norms are equivalent.

Theorem 2.9 ([10, 22]) $\left(\mathcal{L}^\theta(\mu_\infty), ||.||_\theta\right)$ *is a normed linear space when equivalent functions are identified. Moreover* $||f||_\theta \leq 1$ *if and only if* $\int_{\mathcal{B}} \theta(f) d\mu_\infty \leq 1$.

Further Luxemburg norm satisfies monotonicity and strong Fatou property of the function. That is

1. For measureable functions f_1 and f_2 so that $|f_1| \leq |f_2|$ a.e., implies $||f_1||_\theta \leq ||f_2||_\theta$.
2. $0 \leq f_n \uparrow f$ a.e., implies $||f_n||_\theta \nearrow ||f||_\theta \leq \infty$.

Let (θ, ϕ) be a pair of normalized Young functions, i.e., $\theta(1) + \phi(1) = 1$, then for $f \in \mathcal{L}^\theta(\mu_\infty)$ and $g \in \mathcal{L}^\phi(\mu_\infty)$ follows $\int_{\mathcal{B}} |fg| d\mu_\infty \leq ||f||_\theta ||g||_\theta$.

Theorem 2.10 ([10, 22]) $\left(\mathcal{L}^\theta(\mu_\infty), ||.||_\theta\right)$ *is a Banach space if a.e. equal functions are identified. Moreover* $\left(\mathcal{L}^\theta(\mu_\infty), ||_{(\theta)}\right)$ *is a Banach space when* μ_∞*-equivalent functions are identified and* μ_∞ *has the finite subset property.*

We shall recall the modular nature of Orlicz space. In 1950, H. Nakano introduced modular in Orlicz space which is defined as below.

Definition 2.8 ([10, 22]) A functional $\rho_\theta : f \longmapsto \int_{\mathcal{B}} \theta(f) d\mu_\infty$ is called modular. A sequence $\{f_n : n \geq 1\} \subset \mathcal{L}^\theta(\mu_\infty)$ is said to be mean or modular convergent to $f \in \mathcal{L}^\theta(\mu_\infty)$ if $\rho_\theta(f_n - f) \to 0$ as $n \to \infty$. It is strongly or norm convergent to f if $||f_n - f||_\theta \to 0$ as $n \to \infty$.

Theorem 2.11 ([10, 20, 22]) *Let* $\{f_n : n \geq 1\}$ *be a sequence from* $\mathcal{L}^\theta(\mu_\infty)$ *and* $f \in \overline{\mathcal{L}}^\phi(\mu_\infty)$. *Then the following assertions hold:*

1. *The modular (or mean) convergence is implied by the norm (or strong) convergence of* f_n *to* f. *The converse holds if* $\{f_n, f : n \geq 1\} \subset \mathcal{L}^\theta(\mu_\infty)$ *and* θ *is* Δ_2*-regular.*
2. *If* θ *is a* Δ_2*-regular Young function or* ϕ *is continuous and concave* $\phi(0) = 0$, $\phi \nearrow$ *as well,* $\rho_\theta(f_n) \to \rho_\theta(f)$ *as* $n \to \infty$ *and* $f_n \to f$ *a.e., or in* μ_∞*-measure, then* $f_n \to f$ *in norm.*

We shall cover one of the most important property called separability of Orlicz spaces. In order to understand separable property of Orlicz space, one should understand Morse space M^θ, which was introduced by Morse and Transue in 1950.

Definition 2.9 ([20, 22]) The Morse space is defined as follows:

$$M^\theta = \{f \in \mathcal{L}^\theta(\mu_\infty) : kf \in \overline{\mathcal{L}}^\phi(\mu_\infty), \text{ for all } k > 0\}.$$

A step function is a function that takes finitely many finite values, each on a measurable subset of the domain $\mathcal{B}$, and a simple function is a step function which vanishes outside a set of finite μ_∞-measure. Let us consider N^θ is closed linear span of all step functions from $\mathcal{L}^\theta(\mu_\infty)$. Then $M^\theta \subseteq N^\theta \leq \mathcal{L}^\theta(\mu_\infty)$. One should have idea of absolutely continuous norm for further understanding.

Definition 2.10 ([10, 22]) An element $f \in \mathcal{L}^\theta(\mu_\infty)$ is said to have an absolutely continuous norm if $||f\chi_{A_n}||_\theta \to 0$ for each sequence of measurable set $A_n \downarrow \emptyset$.

Remark 2.1 Orlicz space $\mathcal{L}^\theta(\mu_\infty)$ has an absolutely continuous norm if every element of $\mathcal{L}^\theta(\mu_\infty)$ has such a property.

Theorem 2.12 ([22]) *For each $f \in \mathcal{L}^\theta(\mu_\infty)$ the following are mutually equivalent statements:*

1. *f has an absolutely continuous norm.*
2. *For each measurable f_n, $|f| \geq f_n \geq \ldots \to 0$ pointwise, then $||f_n||_\theta \to 0$ as $n \to \infty$.*
3. *For each measurable f_n such that $f_n \to f$ a.e., and $|f_n| \leq |f|$, a.e., we have $||f_n - f||_\theta \to 0$ as $n \to \infty$.*

In addition, we have $\theta(x) = 0$ only for $x = 0$, then M^θ has an absolutely continuous norm and hence $\mathcal{L}^\theta(\mu_\infty)$ has this property if and only if $M^\theta = \overline{\mathcal{L}}^\phi(\mu_\infty)$.

Theorem 2.13 ([10, 22]) *Let θ be a continuous Young function with $\theta(x) = 0$ if and only if $x = 0$. Then*

1. *M^θ is separable if and only if the measure triple $(\mathcal{B}, \Sigma, \mu_\infty)$ is separable.*
2. *$\mathcal{L}^\theta(\mu_\infty)$ is separable if and only if $\mathcal{L}^\theta(\mu_\infty) = M^\theta(\mu_\infty)$ and $(\mathcal{B}, \Sigma, \mu_\infty)$ is separable.*

In particular if θ is Δ_2-regular, then $\mathcal{L}^\theta(\mu_\infty)$ is separable if and only if $(\mathcal{B}, \Sigma, \mu_\infty)$ is separable.

Remark 2.2 Kransnoselskii and Rutickii [10] have explicitly shown that if $(\mathcal{B}, \Sigma, \mu_\infty)$ is a finite diffuse measure and θ is a continuous Young function which does not satisfy Δ_2-condition, then $\mathcal{L}^\theta(\mu_\infty)$ is not separable.

Another important property of Orlicz space is its reflexivity. We recall a Banach space X is reflexive if $X = X^{**}$ where X^* is adjoint of X which is adjoint of X

which is always a Banach space under the adjoint norm. We denote X^{**} be second adjoint of X.

Theorem 2.14 ([10, 22]) *Let $(\mathcal{B}, \Sigma, \mu_\infty)$ be a measure space and (θ, ϕ) be a pair of complementary Young function, then $\mathcal{L}^\theta(\mu_\infty)(\mathcal{L}^\phi(\mu_\infty))$ is reflexive if and only if $\mathcal{L}^\theta(\mu_\infty) = M^\theta(\mu_\infty)$ and $\mathcal{L}^\phi(\mu_\infty) = M^\phi(]mu_\infty)$ or equivalently both $\mathcal{L}^\theta(\mu_\infty)$ and $\mathcal{L}^\phi(\mu_\infty)$ have absolutely continuous norms. In particular if both θ and ϕ are Δ_2-regular.*

Corollary 2.1 *Let $(\mathcal{B}, \Sigma, \mu_\infty)$ be a finite measure space and θ be a continuous Young function if and only if $\theta \in \Delta_2 \cap \nabla_2$.*

Orlicz spaces are discussed in different settings by different researchers (see [18, 19, 21]). The known fact $\mathcal{L}^1(\mu_\infty) \subset \mathcal{L}^1_{loc}(\mu_\infty)$ is used in several settings to discuss many applications of the Orlicz space.

Definition 2.11 ([18, Def 2.1]) Let θ be a convex function, then the Orlicz space can be defined as

$$\mathcal{L}^\theta(\mathbb{R}^n) = \{f \in \mathcal{L}^1_{loc}(\mathbb{R}^n) : ||f||_{\mathcal{L}^\theta(\mathbb{R}^n)} < +\infty\}, \tag{2.3}$$

where for some $\alpha > 0$

$$||f||_{\mathcal{L}^\theta(\mathbb{R}^n)} = \inf\left\{\alpha > 0 : L\int_{\mathbb{R}^n} \theta\left(\frac{|f(x)|}{\alpha}\right) d\mu_\infty(x) \leq 1\right\}.$$

Liu Ning et al. defined weak Orlicz spaces in [21]. We recall the weak Orlicz space as below. Let us consider $(\mathcal{B}, \Sigma, \mu_\infty)$ be a complete measure space with no atom and w be any weight function. Then

$$w\mathcal{L}^\theta(\mu_\infty) = \left\{f : \exists\, \alpha > 0 : \sup_{t>0} \theta\left(\frac{t}{\alpha}\right)\mu_\infty(|f| > t) < \infty\right\}$$

and

$$wM^\theta(\mu_\infty) = \left\{f : \forall\, \alpha > 0 : \sup_{t>0} \theta\left(\frac{t}{\alpha}\right)\mu_\infty(|f| > t) < \infty\right\}$$

It is clear $wM^\theta(\mu_\infty) \subset w\mathcal{L}^\alpha(\mu_\infty)$ and wM^θ is linear. Let

$$||f||_w = \inf\left\{\alpha > 0 : \sup_{t>0} \theta\left(\frac{t}{\alpha}\right)\mu(|f| > t) \leq 1\right\}.$$

The expression $w\mathcal{L}^\theta(\mu_\infty)$ is called weak Orlicz space with $f \in w\mathcal{L}^\theta(\mu_\infty)$ whenever $||f||_w < \infty$. If $\theta(t) = t^p$, then $w\mathcal{L}^\theta(\mu_\infty) = w\mathcal{L}^\theta(\mu_\infty)$, $1 < p < \infty$. As usual $f, g \in w\mathcal{L}^\theta(\mu_\infty)$ whenever $f = g$ a.e. Several results of weak Orlicz spaces were presented in [16, 21].

Theorem 2.15 ([21]) *If $\theta \in \Delta_2$, then $\big(w\mathcal{L}^{\theta}(\mu_\infty), ||.||_w\big)$ is a quasi-norm space.*

2.4 Discrete Orlicz-Morrey Spaces

In this section of the chapter we introduce another version of Orlicz-Morrey space. We discuss completeness of this space in this section. Let $m \in \mathbb{Z}$, $N \in \omega := \mathbb{N}\cup\{0\}$, we write $S_{m,N} := \{m - N, \ldots, m, \ldots, m + N\}$ and $|S_{m,N}| = 2N + 1$ for its cardinality of $S_{m,N}$. Let G_ψ be the set of all functions $\psi : 2\omega + 1 \to (0, \infty)$ such that $\psi(2N + 1)$ is nondecreasing but for any $2M + 1 \in 2\omega + 1$, $\frac{\psi(2(N+M)+1)}{\psi^{-1}(\frac{2(N+M)+1}{2N+1})}$ is nonincreasing.

For $\psi_1, \psi_2 : 2\omega + 1 \to (0, \infty)$, we write $\psi_1 \preceq \psi_2$ if there exists a constant $C > 0$ such that $\psi_1(2N + 1) \le C\psi_2(2N + 1)$ for all $N \in \omega$. If $\psi_1 \preceq \psi_2$ and $\psi_2 \preceq \psi_1$, then we write $\psi_1 \approx \psi_2$.

Now, let θ be a Young function and $\psi \in G_\psi$. The Discrete Orlicz–Morrey space $\ell_{\psi,\theta}(\mathbb{R})$ is the set of all sequences $x = (x_k)_{k=1}^{\infty}$ taking values in $\mathbb{R}$ such that for every $m \in \mathbb{Z}$ and $N \in \omega$, we have

$$\|x\|_{\ell_{\psi,\theta}} := \sup_{m\in\mathbb{Z}, N\in\omega} \psi(2N + 1)\|x\|_{(\theta,m,N)} < \infty,$$

where $\|x\|_{(\theta,m,N)} := \inf\Big\{b > 0 : \frac{1}{|S_{m,N}|} \sum_{k\in S_{m,N}} \theta\Big(\frac{|x_k|}{b}\Big) \le 1\Big\}$.

In the next section, we will show the Discrete Orlicz–Morrey space $\ell_{\psi,\theta}(\mathbb{R})$ is a Banach space with respect to $\|x\|_{\ell_{\psi,\Psi}}$.

For $\psi(2N+1) := 2N+1$, the space $\ell_{\psi,\theta}(\mathbb{R})$ is the Discrete Orlicz space $\ell_\theta(\mathbb{R})$. Meanwhile, for $\theta(t) = t^p$, the space $\ell_{\psi,\theta}(\mathbb{R})$ reduces to the generalized Discrete Morrey space $\ell_\psi^p(\mathbb{R})$.

Throughout this chapter, the letter C denotes a constant that may vary in values from line to line. To keep track of some constants, we use subscripts, such as C_1 and C_2.

2.4.1 Inclusion Properties of Strong Discrete Orlicz–Morrey Spaces

First, we will show $\|\cdot\|_{\ell_{\psi,\Psi}}$ defines a norm on $\ell_{\psi,\theta}(\mathbb{R})$. For getting a result, we present some lemmas in the following.

Lemma 2.1 *If $x = (x_k)_{k=1}^{\infty} \in \ell_{\psi,\Psi}(\mathbb{R})$, then*

$$\frac{1}{|S_{m,N}|} \sum_{k\in S_{m,N}} \theta\bigg(\frac{|x_k|}{\|x\|_{(\theta,m,N)}}\bigg) \le 1$$

for any $m \in \mathbb{Z}$ *and* $N \in \omega$. *Furthermore,* $\|x\|_{(\theta,m,N)} \leq 1$ *if and only if* $\frac{1}{|S_{m,N}|}\sum_{k\in S_{m,N}}\theta(|x_k|) \leq 1$ *for any* $m \in \mathbb{Z}$ *and* $N \in \omega$.

Proof Let x be an element of $\ell_{\psi,\theta}(\mathbb{R})$ and take an arbitrary $\epsilon > 0$, then there exists $b_\epsilon > 0$ such that $b_\epsilon \leq \|x\|_{(\theta,m,N)} + \epsilon$ and $\frac{1}{|S_{m,N}|}\sum\limits_{k\in S_{m,N}}\theta\Big(\frac{|x_k|}{b_\epsilon}\Big) \leq 1$ for any $m \in \mathbb{Z}$ and $N \in \omega$. Because Ψ is increasing, we have

$$\frac{1}{|S_{m,N}|}\sum_{k\in S_{m,N}}\theta\Big(\frac{|x_k|}{\|x\|_{(\theta,m,N)}+\epsilon}\Big) \leq \frac{1}{|S_{m,N}|}\sum_{k\in S_{m,N}}\theta\Big(\frac{|x_k|}{b_\epsilon}\Big) \leq 1.$$

Since $\epsilon > 0$ is arbitrary, we can conclude

$$\frac{1}{|S_{m,N}|}\sum_{k\in S_{m,N}}\theta\Big(\frac{|x_k|}{\|x\|_{(\theta,m,N)}}\Big) \leq 1$$

for any $m \in \mathbb{Z}$ and $N \in \omega$.

Next, if $\|x\|_{(\theta,m,N)} \leq 1$ for any $m \in \mathbb{Z}$ and $N \in \omega$, then

$$\frac{1}{|S_{m,N}|}\sum_{k\in S_{m,N}}\theta(|x_k|) \leq \frac{1}{|S_{m,N}|}\sum_{k\in S_{m,N}}\theta\Big(\frac{|x_k|}{\|x\|_{(\theta,m,N)}}\Big) \leq 1.$$

Now, assume that $\frac{1}{|S_{m,N}|}\sum_{k\in S_{m,N}}\theta(|x_k|) \leq 1$ holds for any $m \in \mathbb{Z}$ and $N \in \omega$. Clearly, we have $\|x\|_{(\theta,m,N)} \leq 1$. □

Lemma 2.2 *Let* θ *be a Young function,* $\psi \in G_\psi$, $m \in \mathbb{Z}$, *and* $N \in \omega$. *Then the following statements are equivalent:*

(1) $\frac{1}{|S_{m,N}|}\sum\limits_{k\in S_{m,N}}\theta\Big(\frac{|x_k|}{\epsilon}\Big) \leq 1$ *for every* $\epsilon > 0$.
(2) $\|x\|_{(\theta,m,N)} = 0$.

Proof Assume that (1) holds. By definition of $\|x\|_{(\theta,m,N)}$ we have $0 \leq \|x\|_{(\theta,m,N)} \leq \epsilon$, for every $\epsilon > 0$. So we conclude that $\|x\|_{(\theta,m,N)} = 0$. Suppose, on the contrary, that there exist $\epsilon_0 > 0$ such that $\frac{1}{|S_{m,N}|}\sum\limits_{k\in S_{m,N}}\theta\Big(\frac{|x_k|}{\epsilon_0}\Big) > 1$. By Lemma 2.1 we have $\|x\|_{(\theta,m,N)} \geq \epsilon_0 > 0$. As a consequence, we conclude that $\frac{1}{|S_{m,N}|}\sum\limits_{k\in S_{m,N}}\theta\Big(\frac{|x_k|}{\epsilon}\Big) \leq 1$ for every $\epsilon > 0$. □

Lemma 2.3 *Let* θ *be a Young function,* $\psi \in G_\psi$, $m \in \mathbb{Z}$, *and* $N \in \omega$. *Then the following statements are equivalent:*

(1) $\frac{1}{|S_{m,N}|}\sum\limits_{k\in S_{m,N}}\theta\Big(\alpha|x_k|\Big) = 0$ *for every* $\alpha > 0$.
(2) $\|x\|_{(\theta,m,N)} = 0$.

Proof Suppose that (1) holds. As before, we can obtain $\|x\|_{(\theta,m,N)} = 0$. Now, suppose that (2) holds. Take an arbitrary $0 < \epsilon \leq 1$. Since θ is a convex function, we have

$$\frac{1}{|S_{m,N}|}\sum_{k\in S_{m,N}}\theta\Big(\alpha|x_k|\Big) \leq \epsilon\left(\frac{1}{|S_{m,N}|}\sum_{k\in S_{m,N}}\theta\Big(\frac{\alpha|x_k|}{\epsilon}\Big)\right) \leq \epsilon.$$

Since $0 < \epsilon \leq 1$ is arbitrary, we can conclude that $\frac{1}{|S_{m,N}|}\sum\limits_{k\in S_{m,N}}\theta\Big(\alpha|x_k|\Big) = 0.$ □

Proposition 2.1 *Let θ be a Young function and $\psi \in G_\psi$, the mapping $\|\cdot\|_{\ell_{\psi,\theta}}$ defines a norm on $\ell_{\psi,\theta}(\mathbb{R})$. Moreover, $(\ell_{\psi,\theta}(\mathbb{R}), \|\cdot\|_{\ell_{\psi,\theta}})$ is a Banach space.*

Proof It is easy to prove that $\|x\|_{\ell_{\psi,\theta}} \geq 0$ and $\|\alpha x\|_{\ell_{\psi,\theta}} = |\alpha|\|x\|_{\ell_{\psi,\theta}}$, for every $x \in \ell_{\psi,\theta}(\mathbb{R})$ and $\alpha \in \mathbb{R}$. Now, we will prove $\|x\|_{\ell_{\psi,\theta}} = 0$ if and only if $x = 0$. If $x = 0$, then we have $\|x\|_{\ell_{\psi,\theta}} = 0$. Let $\|x\|_{\ell_{\psi,\theta}} = 0$, then $\|x\|_{(\theta,m,N)} = 0$ for every $m \in \mathbb{Z}$ and $N \in \omega$. By Lemma 2.3 we have $\frac{1}{|S_{m,N}|}\sum\limits_{k\in S_{m,N}}\theta\Big(\alpha|x_k|\Big) = 0$. In fact, α, $|S_{m,N}|$, and Ψ are positive values, we have $x_k = 0$ for every $k \in S_{m,N}$. Since $m \in \mathbb{Z}$ and $N \in \omega$ are arbitrary, we have $x = 0$.

Next, we will prove $\|x_1 + x_2\|_{\ell_{\psi,\theta}} \leq \|x_1\|_{\ell_{\psi,\theta}} + \|x_2\|_{\ell_{\psi,\theta}}$, for every $x_1, x_2 \in \ell_{\psi,\theta}(\mathbb{R})$. Let $x_1 = (x_k^{(1)})_{k=1}^{\infty}$ and $x_2 = (x_k^{(2)})$ be elements of $\ell_{\psi,\theta}(\mathbb{R})$. For any $m \in \mathbb{Z}$, $N \in \omega$, and $i = 1, 2$, we have

$$\begin{aligned}
&\sum_{k\in S_{m,N}}\theta\Big(\frac{|x_k^{(1)}) + (x_k^{(2)})|}{\|x_1\|_{(\theta,m,N)} + \|x_2\|_{(\theta,m,N)}}\Big)\\
&\leq \sum_{k\in S_{m,N}}\Phi\Big(\sum_{i=1}^{2}\frac{\|x_i\|_{(\theta,m,N)}}{\|x_1\|_{(\theta,m,N)} + \|x_2\|_{(\theta,m,N)}}\frac{|x_k^{(i)}|}{\|x_i\|_{(\theta,m,N)}}\Big)\\
&\leq \sum_{i=1}^{2}\left(\frac{\|x_i\|_{(\theta,m,N)}}{\|x_1\|_{(\theta,m,N)} + \|x_2\|_{(\theta,m,N)}}\sum_{k\in S_{m,N}}\theta\Big(\frac{|x_k^{(i)}|}{\|x_i\|_{(\theta,m,N)}}\Big)\right)\\
&\leq |S_{m,N}|
\end{aligned}$$

or $\frac{1}{|S_{m,N}|}\sum_{k\in S_{m,N}}\theta\Big(\frac{|(x_k^{(1)})+x_k^{(2)})|}{\|x_1\|_{(\theta,m,N)}+\|x_2\|_{(\theta,m,N)}}\Big) \leq 1.$
By definition of $\|x_1 + x_2\|_{(\theta,m,N)}$, we have

$$\|x_1 + x_2\|_{(\theta,m,N)} \leq \|x_1\|_{(\theta,m,N)} + \|x_2\|_{(\theta,m,N)}.$$

Furthermore, we have

$$\begin{aligned}&\psi(2N+1)\|x_1+x_2\|_{(\theta,m,N)}\\&\leq \psi(2N+1)\|x_1\|_{(\theta,m,N)}+\psi(2N+1)\|x_2\|_{(\theta,m,N)}.\end{aligned}$$

By taking supremum over $m \in \mathbb{Z}$ and $N \in \omega$, we conclude that $\|x_1+x_2\|_{\ell_{\psi,\theta}} \leq \|x_1\|_{\ell_{\psi,\theta}} + \|x_2\|_{\ell_{\psi,\theta}}$, for every $x_1, x_2 \in \ell_{\psi,\theta}(\mathbb{R})$. □

Lemma 2.4 ([19, 22]) *Suppose that θ is a Young function and θ^{-1} denotes its inverse, which is given by $\theta^{-1}(s) := \inf\{r \geq 0 : \theta(r) > s\}$ for every $s \geq 0$. Then the following holds:*

(1) $\theta^{-1}(0) = 0$.
(2) $\theta^{-1}(s_1) \leq \theta^{-1}(s_2)$ *for* $s_1 \leq s_2$.
(3) $\theta(\theta^{-1}(s)) \leq s \leq \theta^{-1}(\theta(s))$ *for* $0 \leq s < \infty$.
(4) *If, for some constants $C_1, C_2 > 0$, we have $\theta_2^{-1}(s) \leq C_1\theta_1^{-1}(C_2 s)$, then $\theta_1(\frac{t}{C_1}) \leq C_2\theta_2(t)$ for $t = \theta_2^{-1}(s)$.*

Lemma 2.5 *Let θ be a Young function and $\psi \in G_\psi$. For $m \in \mathbb{Z}$ and $N_0 \in \omega$, let ξ^{m,N_0} be the characteristics sequence given by*

$$\xi^{m,N_0} := \begin{cases} 1, & \text{if } k \in S_{m,N_0} \\ 0, & \text{otherwise} \end{cases}$$

then $\|\xi^{m,N_0}\|_{(\theta,m,N)} = \dfrac{1}{\theta^{-1}\left(\frac{|S_{m,N}|}{|S_{m,N_0}\cap S_{m,N}|}\right)}$.

Proof Let

$$A_{\theta,m,N} := \Big\{b > 0 : \theta\big(\frac{1}{b}\big) \leq \frac{|S_{m,N}|}{|S_{m,N_0}\cap S_{m,N}|}\Big\}$$

and

$$B_{\theta,m,N} := \Big\{r > 0 : \theta(r) > \frac{|S_{m,N}|}{|S_{m,N_0}\cap S_{m,N}|}\Big\}.$$

Observe that,

$$\begin{aligned}\|\xi^{m_0,N_0}\|_{(\theta,m,N)} &:= \inf\Big\{b > 0 : \frac{1}{|S_{m.N}|}\sum_{k\in S_{m,N}} \theta\Big(\frac{|\xi^{m,N_0}|}{b}\Big) \leq 1\Big\}\\ &= \inf\Big\{b > 0 : \frac{|S_{m,N_0}\cap S_{m,N}|}{|S_{m,N}|}\theta\big(\frac{1}{b}\big) \leq 1\Big\}\\ &= \inf\Big\{b > 0 : \theta\big(\frac{1}{b}\big) \leq \frac{|S_{m,N}|}{|S_{m,N_0}\cap S_{m,N}|}\Big\}\\ &= \inf A_{\theta,m,N}.\end{aligned}$$

Meanwhile,

$$\theta^{-1}\Big(\frac{|S_{m,N}|}{|S_{m,N_0}\cap S_{m,N}|}\Big)$$
$$= \inf\Big\{r \geq 0 : \Psi(r) > \frac{|S_{m,N}|}{|S_{m,N_0}\cap S_{m,N}|}\Big\}$$
$$= \inf B_{\theta,m,N}.$$

Let $b = \frac{1}{\theta^{-1}\left(\frac{|S_{m,N}|}{|S_{m,N_0}\cap S_{m,N}|}\right)}$, by Lemma 2.4 (3) we have

$$\theta\Big(\frac{1}{b}\Big) = \theta\left(\theta^{-1}\Big(\frac{|S_{m,N}|}{|S_{m,N_0}\cap S_{m,N}|}\Big)\right) \leq \frac{|S_{m,N}|}{|S_{m,N_0}\cap S_{m,N}|}.$$

By definition of $\|\cdot\|_{(\theta,m,N)}$, we have $\|\xi^{m,N_0}\|_{(\theta,m,N)} \leq \frac{1}{\theta^{-1}\left(\frac{|S_{m,N}|}{|S_{m,N_0}\cap S_{m,N}|}\right)}$.

Suppose, on the contrary, i.e., $\|\xi^{m,N_0}\|_{(\theta,m,N)} < \frac{1}{\theta^{-1}\left(\frac{|S_{m,N}|}{|S_{m,N_0}\cap S_{m,N}|}\right)}$, then $\frac{1}{\|\xi^{m,N_0}\|_{(\theta,m,N)}} > \theta^{-1}\Big(\frac{|S_{m,N}|}{|S_{m,N_0}\cap S_{m,N}|}\Big)$.

By definition of $\theta^{-1}\Big(\frac{|S_{m,N}|}{|S_{m,N_0}\cap S_{m,N}|}\Big)$, there exists a $r_1 \in B_{\theta,m,N}$ such that

$$\frac{1}{\|\xi^{m,N_0}\|_{(\theta,m,N)}} > r_1 \geq \theta^{-1}\Big(\frac{|S_{m,N}|}{|S_{m,N_0}\cap S_{m,N}|}\Big).$$

Since $r_1 \in B_{\theta,m,N}$ we obtain $\frac{1}{r_1} \notin A_{\theta,m,N}$. So we can conclude that $\frac{1}{r_1} \leq \|\xi^{m,N_0}\|_{(\theta,m,N)}$. As a consequence, we obtain $\|\xi^{m,N_0}\|_{(\theta,m,N)} = \frac{1}{\theta^{-1}\left(\frac{|S_{m,N}|}{|S_{m,N_0}\cap S_{m,N}|}\right)}$, as desired. □

Proposition 2.2 *Let θ be a Young function and $\phi \in G_\psi$. For $m \in \mathbb{Z}$ and $N_0 \in \omega$, let ξ^{m,N_0} be the characteristics sequence given by*

$$\xi^{m,N_0} := \begin{cases} 1, & if\, k \in S_{m,N_0} \\ 0, & otherwise \end{cases}$$

then we have $\|\xi^{m,N_0}\|_{\ell_{\psi,\theta}} = \frac{\psi(2N_0+1)}{\theta^{-1}(1)}$.

Proof Take arbitrary $m \in \mathbb{Z}$ and $N_0 \in \omega$. By Lemma 2.5 we have

$$\|\xi^{m,N_0}\|_{\ell_{\psi,\theta}} = \sup_{m\in\mathbb{Z},N\in\omega} \|\xi^{m,N_0}\|_{(\theta,m,N)}$$

$$= \sup_{m\in\mathbb{Z},N\in\omega} \frac{\psi(2N+1)}{\theta^{-1}\left(\frac{|S_{m,N}|}{|S_{m,N_0}\cap S_{m,N}|}\right)}$$

$$\geq \frac{\psi(2N_0+1)}{\theta^{-1}(1)}.$$

Now, we will prove $\|\xi^{m,N_0}\|_{\ell_{\psi,\theta}} \leq \frac{\psi(2N_0+1)}{\theta^{-1}(1)}$.

Case I: For $N \geq N_0$, write $N = N_0 + N_1$ for $N_1 \in \omega$. Observe that,

$$\frac{\psi(2N_0+1)}{\theta^{-1}(1)} = \frac{\psi((2N_0+1)+0)}{\theta^{-1}\left(\frac{(2N_0+1)+0}{2N_0+1}\right)}$$

$$\geq \frac{\psi(2(N_0+N_1)+1)}{\theta^{-1}\left(\frac{2(N_0+N_1)+1}{2N_0+1}\right)}$$

$$= \frac{\psi(2N+1)}{\theta^{-1}\left(\frac{2N+1}{2N_0+1}\right)}$$

$$= \frac{\psi(2N+1)}{\theta^{-1}\left(\frac{|S_{m,N}|}{|S_{m,N_0}\cap S_{m,N}|}\right)}.$$

So we get, $\frac{\psi(2N+1)}{\theta^{-1}\left(\frac{|S_{m,N}|}{|S_{m,N_0}\cap S_{m,N}|}\right)} \leq \frac{\psi(2N_0+1)}{\theta^{-1}(1)}$.

Case II: For $N < N_0$, we have $\psi(2N+1) \leq \psi(2N_0+1)$. It shows that

$$\frac{\psi(2N+1)}{\theta^{-1}\left(\frac{|S_{m,N}|}{|S_{m,N_0}\cap S_{m,N}|}\right)} = \frac{\psi(2N+1)}{\theta^{-1}\left(\frac{2N+1}{2N+1}\right)} \leq \frac{\psi(2N_0+1)}{\theta^{-1}(1)}.$$

Since, Case I and II are true for arbitrary $m \in \mathbb{Z}$, $N \in \omega$, we have

$$\|\xi^{m,N_0}\|_{\ell_{\phi,\Phi}} = \sup_{m\in\mathbb{Z},N\in\omega} \|\xi^{m,N_0}\|_{(\theta,m,N)}$$

$$= \sup_{m\in\mathbb{Z},N\in\omega} \frac{\psi(2N+1)}{\theta^{-1}\left(\frac{|S_{m,N}|}{|S_{m,N_0}\cap S_{m,N}|}\right)}$$

$$\leq \frac{\psi(2N_0+1)}{\theta^{-1}(1)},$$

as desired. □

Now we come to inclusion properties of (strong) Discrete Orlicz–Morrey spaces in the following.

Theorem 2.16 *Let θ_1, θ_2 be Young functions such that $\theta_1 \prec \theta_2$ and $\psi_1, \psi_2 \in G_\psi$. Then the following statements are equivalent:*

(1) $\psi_1 \preceq \psi_2$ *(on $2\omega + 1$).*
(2) $\ell_{\psi_2,\theta_2}(\mathbb{R}) \subseteq \ell_{\psi_1,\theta_1}(\mathbb{R})$.
(3) *There exists a constant $C > 0$ such that*

$$\|x\|_{\ell_{\psi_1,\theta_1}} \leq C\|x\|_{\ell_{\psi_2,\theta_2}},$$

for every $x \in \ell_{\psi_2,\theta_2}(\mathbb{R})$.

Proof Let us first prove that (1) implies (2). Let $x \in \ell_{\psi_2,\theta_2}(\mathbb{R})$. Recall that $\theta_1 \prec \theta_2$ and $\psi_1 \preceq \psi_2$ means that there exists constant $C_1, C_2 > 0$ such that $\theta_1(t) \leq \theta_2(C_1 t)$ every $t > 0$ and $\psi_1(2N+1) \leq C_2\psi_2(2N+1)$ for every $N \in \omega$. For every $m \in \mathbb{Z}$ and $N \in \omega$, we have

$$\begin{aligned}\frac{1}{|S_{m,N}|}\sum_{k\in S_{m,N}}\theta_1\Big(\frac{|x_k|}{C_1\|x\|_{(\theta_2,m,N)}}\Big) &\leq \frac{1}{|S_{m,N}|}\sum_{k\in S_{m,N}}\theta_2\Big(\frac{C_1|x_k|}{C_1\|x\|_{(\theta_2,m,N)}}\Big)\\ &= \frac{1}{|S_{m,N}|}\sum_{k\in S_{m,N}}\theta_2\Big(\frac{|x_k|}{\|x\|_{(\theta_2,m,N)}}\Big)\\ &\leq 1.\end{aligned}$$

By definition of $\|x\|_{(\theta_1,m,N)}$, we have $\|x\|_{(\theta_1,m,N)} \leq C_2\|x\|_{(\theta_2,m,N)}$. Furthermore, we have

$$\begin{aligned}\|x\|_{\ell_{\psi_1,\theta_1}} &:= \sup_{m\in\mathbb{Z},N\in\omega}\psi_1(2N+1)\|x\|_{(\theta_1,m,N)}\\ &\leq \sup_{m\in\mathbb{Z},N\in\omega}C_1C_2\psi_2(2N+1)\|x\|_{(\theta_2,m,N)}\\ &= C_1C_2\|x\|_{\ell_{\psi_2,\theta_2}}.\end{aligned}$$

This proves that $\ell_{\psi_2,\theta_2}(\mathbb{R}) \subseteq \ell_{\psi_1,\theta_1}(\mathbb{R})$.

Next, since $(\ell_{\psi_2,\theta_2}(\mathbb{R}), \ell_{\psi_1,\theta_1}(\mathbb{R}))$ is a Banach pair, it follows from [11, Lemma 3.3] that (2) and (3) are equivalent. It thus remains to show that (3) implies (1).

Assume that (3) holds. Let $m_0 \in \mathbb{Z}$ and $N \in \omega$. By Proposition 2.2, we have

$$\frac{\psi_1(2N_0+1)}{\theta_1^{-1}(1)} = \|\xi^{m,N_0}\|_{\ell_{\psi_1,\theta_1}} \leq C\|\xi^{m,N_0}\|_{\ell_{\psi_2,\theta_2}} = \frac{C\psi_2(2N_0+1)}{\theta_2^{-1}(1)},$$

whence $\psi_1(2N_0+1) \leq \frac{C\theta_1^{-1}(1)}{\theta_2^{-1}(1)}\psi_2(2N_0+1)$. Since $N_0 \in \omega$ is arbitrary, we conclude that $\psi_1(2N+1) \leq C_1\psi_2(2N+1)$ for every $N \in \omega$, where $C_1 = \frac{C\theta_1^{-1}(1)}{\Psi_2^{-1}(1)}$. □

2.4.2 Inclusion Properties of Weak Discrete Orlicz–Morrey Spaces

We have discussed inclusion properties of Discrete Orlicz–Morrey spaces. Now we discuss inclusion properties of weak Discrete Orlicz–Morrey spaces. First we give definition of weak Discrete Orlicz–Morrey spaces and some lemmas in the following.

Let θ be a Young function and $\psi \in G_\psi$. The weak Discrete Orlicz–Morrey space $w\ell_{\psi,\theta}(\mathbb{R})$ is the set of all sequences $x = (x_k)_{k=1}^{\infty}$ taking values in $\mathbb{R}$ such that

$$\|x\|_{w\ell_{\psi,\theta}} := \sup_{m\in\mathbb{Z}, N\in\omega} \psi(2N+1)\|x\|^{w}_{(\theta,m,N)} < \infty,$$

where

$$\|x\|^{w}_{(\theta,m,N)} := \inf\left\{b > 0 : \sup_{t>0} \frac{\theta(t)\left|k \in S_{m,N} : \frac{|x_k|}{b} > t\}\right|}{|S_{m,N}|} \leq 1\right\}$$

is finite.

In the next section, we will show the weak Discrete Orlicz–Morrey space $w\ell_{\psi,\theta}(\mathbb{R})$ is a quasi Banach space with respect to $\|x\|_{w\ell_{\psi,\theta}}$.

For $\psi(2N+1) := 2N+1$, the space $w\ell_{\psi,\theta}(\mathbb{R})$ is the weak Discrete Orlicz space $w\ell_{\theta}(\mathbb{R})$. Meanwhile, for $\theta(t) = t^p$, the space $w\ell_{\psi,\theta}(\mathbb{R})$ reduces to the generalized Discrete Morrey space $w\ell^{p}_{\psi}(\mathbb{R})$.

The relation between $w\ell_{\psi,\theta}(\mathbb{R})$ and $\ell_{\psi,\theta}(\mathbb{R})$ is presented in the following lemma. (We leave the proof to the reader.)

Lemma 2.6 *Let θ be a Young function and $\psi \in G_\psi$. Then $\ell_{\psi,\theta}(\mathbb{R}) \subseteq w\ell_{\psi,\theta}(\mathbb{R})$ with*

$$\|x\|_{w\ell_{\psi,\theta}(\mathbb{R})} \leq \|x\|_{\ell_{\psi,\theta}(\mathbb{R})}$$

for every $x \in \ell_{\psi,\theta}(\mathbb{R})$.

The following is an analog of Lemma 2.4.

Lemma 2.7 *Let θ be a Young function and $\psi \in G_\psi$. For $m \in \mathbb{Z}$ and $N_0 \in \omega$, let ξ^{m,N_0} be the characteristics sequence given by*

$$\xi^{m,N_0} := \begin{cases} 1, & \text{if } k \in S_{m,N_0} \\ 0, & \text{otherwise} \end{cases}$$

then $\|\xi^{m,N_0}\|^{w}_{(\theta,m,N)} = \dfrac{1}{\theta^{-1}\left(\frac{|S_{m,N}|}{|S_{m,N_0} \cap S_{m,N}|}\right)}$.

The following proposition indicates that the characteristic sequence is also obtained in Discrete weak Orlicz–Morrey spaces.

Proposition 2.3 *Let θ be a Young function and $\psi \in G_{\psi}$. For $m \in \mathbb{Z}$ and $N_0 \in \omega$, let ξ^{m,N_0} be the characteristics sequence given by*

$$\xi^{m,N_0} := \begin{cases} 1, & \text{if } k \in S_{m,N_0} \\ 0, & \text{otherwise} \end{cases}$$

then $\|\xi^{m,N_0}\|_{w\ell_{\psi,\theta}} = \frac{\psi(2N_0+1)}{\theta^{-1}(1)}$.

Proof Take arbitrary $m \in \mathbb{Z}$ and $N_0 \in \omega$. By Lemma 2.7 we have

$$\begin{aligned} \|\xi^{m,N_0}\|_{w\ell_{\psi,\theta}} &= \sup_{m\in\mathbb{Z}, N\in\omega} \|\xi^{m,N_0}\|^{w}_{(\theta,m,N)} \\ &= \sup_{m\in\mathbb{Z}, N\in\omega} \frac{\psi(2N+1)}{\theta^{-1}\left(\frac{|S_{m,N}|}{|S_{m,N_0}\cap S_{m,N}|}\right)} \\ &\geq \frac{\psi(2N_0+1)}{\theta^{-1}(1)}. \end{aligned}$$

On the other hand, by Proposition 2.2 and Lemma 2.6 we have $\|\xi^{m_0,N_0}\|_{w\ell_{\psi,\theta}} \leq \frac{\psi(2N_0+1)}{\theta^{-1}(1)}$, as desired. □

Now we come to inclusion properties of weak Discrete Orlicz–Morrey spaces in the following.

Theorem 2.17 *Let θ_1, θ_2 be Young functions such that $\theta_1 \prec \theta_2$ and $\psi_1, \psi_2 \in G_{\psi}$. Then the following statements are equivalent:*

(1) *$\psi_1 \preceq \psi_2$ (on $2\omega + 1$).*
(2) *$w\ell_{\psi_2,\Psi_2}(\mathbb{R}) \subseteq w\ell_{\psi_1,\theta_1}(\mathbb{R})$.*
(3) *There exists a constant $C > 0$ such that*

$$\|x\|_{w\ell_{\psi_1,\theta_1}} \leq C\|x\|_{w\ell_{\psi_2,\theta_2}},$$

for every $x \in w\ell_{\psi_2,\theta_2}(\mathbb{R})$.

Proof Let us first prove that (1) implies (2). Let $x \in \ell_{\psi_2,\theta_2}(\mathbb{R})$. Recall that $\theta_1 \prec \theta_2$ and $\psi_1 \preceq \psi_2$ means that there exists constant $C_1, C_2 > 0$ such that $\theta_1(t) \leq \theta_2(C_1 t)$ for every $t > 0$ and $\psi_1(2N+1) \leq C_2\psi_2(2N+1)$ for every $N \in \omega$. For $m \in \mathbb{Z}$ and $N \in \omega$, let

$$A_{\theta_1,M,N} = \left\{ b > 0 : \sup_{t>0} \frac{\theta_1(t)\left|\{k \in S_{m,N} : \frac{|x_k|}{b} > t\}\right|}{|S_{m,N}|} \leq 1 \right\}$$

and

$$A_{\theta_2,m,N} = \left\{ b > 0 : \sup_{t>0} \frac{\theta_2(C_1 t)\left|\{k \in S_{m,N} : \frac{|x_k|}{b} > t\}\right|}{|S_{m,N}|} \leq 1 \right\}$$
$$= \left\{ b > 0 : \sup_{t_1>0} \frac{\theta_2(t_1)\left|\{x \in S_{m,N} : \frac{|x_k|}{b} > \frac{t_1}{C_1}\}\right|}{|S_{m,N}|} \leq 1 \right\}$$
$$= \left\{ b > 0 : \sup_{t_1>0} \frac{\theta_2(t_1)\left|\{k \in S_{m,N} : \frac{|C_1 x_k|}{b} > t_1\}\right|}{|S_{m,N}|} \leq 1 \right\}.$$

Then $\|C_1 x\|^{w}_{(\theta_2,m,N)} = \inf A_{\theta_2,m,N}$. Observe that, for arbitrary $b \in A_{\theta_2,m,N}$ and $t > 0$, we have (by setting $t_1 = C_1 t$)

$$\frac{\theta_1(t)\left|\{k \in S_{m,N} : \frac{|x_k|}{b} > t\}\right|}{|S_{m,N}|} \leq \frac{\theta_2(C_1 t)\left|\{k \in S_{m,N} : \frac{|x_k|}{b} > t\}\right|}{|S_{m,N}|}$$
$$= \frac{\theta_2(t_1)\left|\{k \in S_{m,N} : \frac{|C_1 x_k|}{b} > t_1\}\right|}{|S_{m,N}|}$$
$$\leq \sup_{t_1>0} \frac{\theta_2(t_1)\left|\{k \in S_{m,N} : \frac{|C_1 x_k|}{b} > t_1\}\right|}{|S_{m,N}|}$$
$$\leq 1.$$

Since $t > 0$ is arbitrary, we have $\sup\limits_{t>0} \frac{\theta_1(t)\left|\left\{k \in S_{m,N} : \frac{|x_k|}{b} > t\right\}\right|}{|S_{m,N}|} \leq 1$. Hence it follows that $b \in A_{\theta_1,m,N}$, and so we conclude that $A_{\theta_2,m,N} \subseteq A_{\theta_1,m,N}$. Accordingly, we obtain

$$\|x\|^{w}_{(\theta_1,m,N)} = \inf A_{\theta_1,m,N} \leq \inf A_{\theta_2,m,N} = \|C_1 x\|^{w}_{(\theta_2,m,N)} = C_1 \|x\|^{w}_{(\theta_2,m,N)}.$$

Furthermore, we have

$$\|x\|_{w\ell_{\psi_1,\theta_1}} := \sup_{m\in\mathbb{Z}, N\in\omega} \psi_1(2N+1)\|x\|^{w}_{(\theta_1,m,N)}$$
$$\leq \sup_{m\in\mathbb{Z}, N\in\omega} C_1 C_2 \psi_2(2N+1)\|x\|^{w}_{(\theta_2,m,N)}$$
$$= C_1 C_2 \|x\|_{w\ell_{\psi_2,\theta_2}}.$$

This proves that $w\ell_{\psi_2,\theta_2}(\mathbb{R}) \subseteq w\ell_{\psi_1,\theta_1}(\mathbb{R})$.

As mentioned in [18, Appendix G], we know that Lemma 3.3 in [11] still holds for quasi-Banach spaces, so (2) and (3) are equivalent.

Now, we will show that (3) implies (1). Assume that (3) holds. Let $m \in \mathbb{Z}$ and $N_0 \in \omega$. By Proposition 2.3, we have

$$\frac{\psi_1(2N_0+1)}{\theta_1^{-1}(1)} = \|\xi^{m,N_0}\|_{w\ell_{\psi_1,\theta_1}} \leq C\|\xi^{m,N_0}\|_{w\ell_{\psi_2,\theta_2}} = \frac{C\psi_2(2N_0+1)}{\theta_2^{-1}(1)},$$

whence $\psi_1(2N_0+1) \leq \frac{C\theta_1^{-1}(1)}{\theta_2^{-1}(1)}\psi_2(2N_0+1)$. Since $N_0 \in \omega$ is arbitrary, we conclude that $\psi_1(2N+1) \leq C_1\psi_2(2N+1)$ for every $N \in \omega$, where $C_1 = \frac{C\theta_1^{-1}(1)}{\theta_2^{-1}(1)}$. □

References

1. A-Algwaiz, M.: Theory of Distributions. Pure and Applied Mathematics, Monograph. Marcel Dekker (1992)
2. Alexiewicz, A.: Linear functionals on Denjoy-integrable functions. Collectanea Mathematica **1**, 289–293 (1948)
3. Bardaro, C., Vinti, G.: Modular convergence theorems in fractional Musielak-Orlicz spaces. Zeitschrift für Analysis und ihre Anwendungen. J. Anal. Appl. **13**(1), 155–170 (1994)
4. Donaldson, T.K., Trudinger, N.S.: Orlicz Sobolev space and imbedding theorems. J. Funct. Anal. **8**, 52–75 (1971)
5. Kaminska, A., Hudzik, H.: Some remarks on convergence in Orlicz space. Annales Societatis Mathematicae Polonae, Series I: Commentationes Mathematicae, XXI (1979)
6. Kaminska, A.: The criteria for local Uniformly rotundity of Orlicz spaces. Studia Mathematica **79**(3), 201–215 (1984)
7. Kaminska, A.: On Uniform convexity of Orlicz spaces. Indagationes Mathematicae (Proceedings) **85**(1), 27–36 (1982)
8. Kantorovich, L.A., Akilov, G.P.: Functional Analysis, 2nd edn. Pergamon Press, Oxford (1982)
9. Kozek, A.: Orlicz spaces of functions with values in Banach spaces. Commun. Math. **19**, 259–288 (1976)
10. Krasonsel'skii, M.A., Rutickii, Ya.B.: Convex Functions and Orlicz Spaces. P. Noordfoff, Gronin-gen (1961)
11. Krein, S.G., Petunia, Yu.I., Semënov, E.M.: Interpolation of Linear Operators, Translation of Mathematical Monograph 54. American Mathematical Society, Providence (1982)
12. Lindenstrauss, J., Tzafriri, L.: Classical Banach Spaces II, Function Spaces. A Series of Modern Surveys in Mathematics. Springer, Berlin-Heidelberg-New York (1979)
13. Liu, P., Wang, M.: Weak Orlicz spaces: Some basic properties and their applications to harmonic analysis. Sci. China Math. **56**(4), 789–802 (2013)
14. Luxemburg, W.A.J.: Banach Function Spaces, Thesis, Delft (1955)
15. Maligranda, L.: Orlicz spaces and Interpolation, Seminarios De Matematica, Departmento de Matematica, Universidade Estadual de Campinas (1989)
16. Masta, A., Gunawan, A.H., Budhi, W.S.: Inclusion properties of Orlicz and weak Orlicz spaces. J. Math. Fund. Sci. **48**(3), 193–203 (2016)
17. Musielak, J., Orlicz, W.: On modular spaces. Studia Mathematica **18**, 49–65 (1959)
18. Nakai, E.: Orlicz-Morrey Spaces and Some Integral Operators. Kyoto University Research Repository, E. (2004)
19. Nakai, E.: On generalized fractional integrals. Taiwan. J. Math. **5**(3), 587–602 (2001)
20. Nakano, H.: Generalized modular spaces. Studia Mathematica **31**(5), 439–449 (1968)

21. Ning, L. Yonggang, Y.: Weak Orlicz space and its convergence theorems. Acta Mathematica Scientia **30B**(5), 1492–1500 (2010)
22. Rao, M.M., Ren, Z.: The Theory of Orlicz Spaces. Marcel Dekker, New Work (1991)
23. Roy, S.K., Chakraborty, N.D.: Orlicz spaces for a family of measures. Analysis Mathematica **12**, 229–235 (1986)
24. Roy, S.K., Chakraborty, N.D.: On Orlicz spaces for a family of measures. J. Math. Anal. Appl. **145**(2), 485–503 (1990)
25. Rubshtein, B.Z.A., Grabarnik, G.Y., Muratov, M.A., Pashkova, Y.S.: Foundations of Symmetric Spaces of Measurable Functions. Developments in Mathematics. Springer, Berlin (2016)
26. Skaff, M.S.: Vector valued Orlicz spaces generalized N-function, I. Pac. J. Math. **28**(1), 193–206 (1969)
27. Skaff, M.S.: Vector valued Orlicz spaces, II. Pac. J. Math. **28**(2), 1–18 (1969)
28. Thung, T.V.: Some collections of functions Dense in an Orlicz Space. Acta Mathematica Vietnamica **25**(2), 195–208 (2000)

Chapter 3
Historical Background of Non-absolute Integrals

Differential and integral calculus were independently developed at the end of the seventeenth century by Isaac Newton (1642–1727) and Gottfried Leibnitz (1646–1716). A curve's slope at a certain location was determined using differential calculus, and the area under a curve was calculated using integral calculus. It turns out that there is an inverse relationship between the differential and integral calculus for a wide class of functions, meaning that a differentiable function can be derived up to a constant by integrating its derivative. Mathematicians started to understand during the seventeenth century that the very base of mathematical analysis was highly unstable and vaguely defined. This acted as a major barrier to study once geometric intuition was no longer sufficient. In the nineteenth century, as a result, the modern treatment of continuity is typically attributed to Augustin-Louis Cauchy (1789–1857), but there are many more mathematicians who made important contributions in the direction as well, see [30]. During this time integration, theorists began to acquire tools that would allow them to rigorously define and study various integrals. In 1854 Bernhard Riemann (1826–1866) introduced the Riemann integral which was one of the first formally defined integrals. In 1904 Henri Lebesgue (1875–1941) was able to show that a real-valued function defined on a compact interval is Riemann integrable if and only if it is bounded and discontinuous on a set of Lebesgue measure 0. There are bounded derivatives that are discontinuous on a set of positive Lebesgue measures, and thus such derivatives are not Riemann integrable. An example of such a function was given by Vito–Volterra (1860–1940) in 1981 as follows:

$$f_a(x) = \begin{cases} (x-a)^2 \sin \frac{1}{x-a} \text{ if } x \neq a, \\ 0 \text{ if } x = a. \end{cases}$$

This drawback was remedied by the Lebesgue integral which Lebesgue introduced in [38]. This fact gives the Lebesgue integral is a generalization of the Riemann

B. Hazarika, H. Kalita, *Theory of Henstock–Orlicz Spaces*, Mathematical Marvels: Texts and Monographs in the Spirit of CR Rao,
https://doi.org/10.1007/978-981-96-9548-5_3

integral, and it can integrate all bounded derivatives restricted to compact intervals. When $F' = f$ on $[a, b] \setminus S$, if $[a, b]$ is compact, then by Lebesgue's fundamental theorem of calculus we have that f is Lebesgue integrable on $[a, b]$ and F can be re-constructed by the Lebesgue integral if and only if F is absolute continuity, thus it is easy to show that there are differentiable functions which cannot be recovered from their respective derivatives via the Lebesgue integral. Unfortunately, Lebesgue integral was not understood by mathematicians during that period. Measure theory was a very new topic, and functional analysis had not yet been developed to a point where these two areas of mathematics could be researched in unity to produce a more complete theory. During the twentieth century mathematicians incrementally developed the necessary tools to study the Lebesgue integral more thoroughly, and Lebesgue integral became the standard integral in advanced mathematical analysis. In 1912 Arnaud Denjoy (1884–1974) presented a powerful integral which was able to integrate all finite derivatives and recover their primitive functions which were not possible in the case of Lebesgue integral. In [36, Section 1.1] it is mentioned Denjoy constructed a transfinite sequence $(I_k)_{k \leq \Omega}$ of increasingly general integrals, where Ω is the first uncountable ordinal, I_0 is the Lebesgue integral, and I_Ω is so-called narrow Denjoy integral. In 1914 Oskar Perron (1880–1975) introduced an integral, which after some modification in 1915 by Hans Bauer (1891–1953) also able to integrate all finite derivatives and recover their primitive functions. After a few years, mathematicians had realized that the narrow Denjoy integral and the Perron integral are equivalent with the so-called Hake-Aleksandrov-Looman theorem.

3.1 Henstock-Kurzweil Integral

In 1957, Jaroslav-Kurzweil (1926–2022) published a paper on differential equations in which he discussed a new integral. Four years later, in 1961, while unaware of the work of Kurzweil, Ralph Henstock (1923–2007) published a paper on integration theory in which he discussed the same integral as Kurzweil. Throughout a series of papers in the 1960s, Henstock developed a substantial amount of properties of this integral. The definition of this integral as defined by Kurzweil in [37] and Henstock in [35] is quite elegant as it is highly reminiscent of the Riemann integral and since a substantial amount of its properties can be developed using Riemann sums and basic epsilon-delta proofs. In honor of these mathematicians, nowadays this integral is called the Henstock-Kurzweil integral. During the late 1990s a lot of integration theorists had researched the Henstock-Kurzweil integral extensively, and consequently the theory of this integral had been highly refined. For details of this integral, one can see [1, 31, 43]. Recalling, Henstock-Kurzweil integral was first developed by R. Henstock and J. Kurzweil independently, during 1957–1961 from Riemann integral with the concept of tagged partitions and gauge functions. Henstock-Kurzweil integral (we write HK-integral) is a kind of non-absolute integral and contains Lebesgue integral, see [31, 32, 37]. One can see

[2–24, 26–29, 31–37, 41–43] for details of Henstock-Kurzweil and other related integrals in different settings.

It is known that if f is bounded with compact support, then the following are equivalent:

(a) f is Henstock-Kurzweil integrable.
(b) f is Lebesgue integrable.
(c) f is Lebesgue measurable.

In general, every Henstock-Kurzweil integrable function is measurable, and f is Lebesgue integrable if and only if both f and $|f|$ are Henstock-Kurzweil integrable. This means that the Henstock-Kurzweil integral can be thought of as a "non-absolutely convergent version of Lebesgue integral."

Recall, a tagged partition of an interval $I = [a, b]$ is a finite set of ordered pairs $D = \{(t_i, I_i) : 1 \leq i \leq m\}$, where $\{I_i : 1 \leq i \leq m\}$ is a partition of I consisting of closed nonoverlapping subintervals and t_i is a point in I_i; t_i is called the tag associated with I_i. If $f : I \to \mathbb{R}$, the Riemann sum of f with respect to D is defined as

$$S(f, D) = \sum_{i=1}^{m} f(t_i)l(I_i), \tag{3.1}$$

where $l(I_i)$ is the length of the subinterval I_i.

If $\delta : I \to (0, \infty)$ is a positive function, we define an open interval valued function on I by setting $\gamma(t) = (t - \delta(t), t + \delta(t))$. If $I_i = [x_i, x_{i+1}]$, and $t_i \in I_i$, $1 \leq i \leq m$, we can rewrite

$$t_i \in I_i \subset \gamma(t_i). \tag{3.2}$$

A function γ defined on I such that $\gamma(f)$ is an open interval containing t for each $t \in I$ is called a gauge on I. If $D = \{(t_i, I_i) : 1 \leq i \leq m\}$ is a tagged partition of I and γ is a gauge on I, we say that D is δ-fine if (3.2) is satisfied, i.e., if D is δ-fine, we write $D << \delta$.

Definition 3.1 ([43, Definition 2]) A function $f : [a, b] \to \mathbb{R}$ is said to be Henstock-Kurzweil integrable over $[a, b]$ if there exists $A \in \mathbb{R}$ such that for every $\epsilon > 0$ there exists a gauge $\delta : [a, b] \to \mathbb{R}$ such that $|S(f, D) - A| < \epsilon$ whenever D is a δ-fine tagged partition of $[a, b]$. The number A is called the Henstock-Kurzweil integral of f over $[a, b]$ and is denoted by $HK \int_a^b f$.

The space of Henstock-Kurzweil integrable functions is a vector space under the operations of pointwise addition and scalar multiplication, and the space also has a natural semi-norm, called Alexiewicz norm. When $I = [a, b]$, $-\infty < a < b < \infty$.

Let $HK(I)$ be the space of all Henstock-Kurzweil integrable functions defined on I. The natural semi-norm on $HK(I)$ of Alexiewicz is

$$||f|| = \sup\left\{\left|\int_a^t f\right| : a \leq t \leq b\right\}. \tag{3.3}$$

In contrast to the case of the space of Lebesgue integrable functions with L^1-norm, the space $HK(I)$ is not complete under the norm (3.3). Interestingly if $f = 0$ a.e. in I, with the Theorem 2 of Appendix 3 of [43], $HK(I)$ is normed space under (3.3) norm. There are several ways to approach the process of recovering a function from its derivative in $\mathbb{R}^n$, which defines the Henstock-Kurzweil integral on Euclidean spaces. Tuo-Yeong-Lee [44] discussed in detail. All norms are equivalent on $\mathbb{R}^n$; however, for the Henstock-Kurzweil integral the maximal norm $||x|| = \max_{1\leq k\leq n} |x_k|$ is natural.

Henstock-Integral on $\mathbb{R}^n$ Let $\mathbb{R}^n$ be n-dimensional Euclidean space. A typical element of $\mathbb{R}^n$ will be denoted by $x = (x_1, x_2, \ldots, x_n)$. An interval in $\mathbb{R}^n$ is a set of the form $J = [\mathbf{a}, \mathbf{b}] := \prod_{i=1}^n [a_i, b_i]$, where $-\infty < a_i < b_i < \infty$ for $i = 1, 2 \ldots, n$.. The set $\prod_i^n [a_i, b_i] \subset \mathbb{R}^n$ is known as a degenerate interval if $a_i = b_i$ for some $i \in \{1, 2, \ldots, n\}$. Two intervals $J = [\mathbf{a}, \mathbf{b}]$, $K = [\mathbf{u}, \mathbf{v}]$ in $\mathbb{R}^n$ are said to be nonoverlapping if $\prod_i^n (a_i, b_i) \cap \prod_i^n (u_i, v_i)$ is empty. Also, union of two intervals in $\mathbb{R}^n$ to be an interval in $\mathbb{R}^n$, see [44, Lemma 2.1.2]. We know that the space $\mathbb{R}^n$ equipped with the maximum norm $||.||$, where $||x|| = \max_{1\leq k\leq n} |x_k|$. With this norm, we denote the closed ball of $\mathbb{R}^n$ by $B[\mathbf{x}, r] = \{\mathbf{x} \in \mathbb{R}^n : ||\mathbf{y} - \mathbf{x}|| \leq r\}$, whose center is $\mathbf{x}$ with sides parallel to the coordinate axes of length $2r$. It is a closed interval for side i, about x_i, such that $x_i[a_i, b_i]$. So, let $B[\mathbf{x}, r] = [J, \mathbf{x}]$, where $J = \prod_{i=1}^n [a_i, b_i]$, J is closed interval in $\mathbb{R}^n$.

Definition 3.2 If E is a compact ball in $\mathbb{R}^n$, a partition P of E is a collection $\{(J_i, x_i) : x_i \in J_i, 1 \leq i \leq m\}$, where $J_1, J_2, \ldots, J_m$ are nonoverlapping closed intervals, i.e., $\mu_\infty[J_i \cap J_j] = 0, i \neq j$ and $\bigcup_{i=1}^m J_i = E$.

If δ is a positive function on E, we say P is Henstock δ-partition of E if for each i, $J_i \subset B'(x_i, \delta(x_i))$. The function δ is a gauge on E.

Definition 3.3 A function $f : E \to \mathbb{R}$ is said to be Henstock-Kurzweil (in short HK) integrable on E, if there exists a number A such that for any $\epsilon > 0$ there exist a gauge δ and Henstock δ-partition on E such that

$$\left|\sum_{i=1}^m f(x_i)\mu_\infty(J_i) - A\right| < \epsilon.$$

Definition 3.4 A measurable function $f : \mathbb{G} \subset \mathbb{R}^n \to \mathbb{R}$ is called locally Henstock-Kurzweil integrable if $f\chi_{\mathcal{K}} \in HK(\mathcal{K})$ for all $\mathcal{K} \subseteq \mathbb{G}$ compact. We denote the set of locally Henstock-Kurzweil integrable functions as HK_{loc}.

Recalling that the function f is Henstock integrable on a measurable set $\mathbb{G} \subset \mathbb{R}^n$ if $f\chi_{\mathbb{G}}$ is Henstock-Kurzweil integrable on $\mathbb{G}$. That is, $f \in HK_{loc}(\mathbb{R}^n)$ means $f \in HK(\mathcal{K})$, where $\mathcal{K} \subseteq \mathbb{G}$ compact. Also with easy analogous, $L^1_{loc}(\mathbb{R}^n) \subset HK_{loc}(\mathbb{R}^n)$.

3.1.1 *Henstock-Kurzweil (HK) Integral with Lebesgue Measure*

There are two ways to define an integral. One can provide a descriptive definition or an operational (or constructive). A descriptive definition describes the integral in relationship to its derivative without proving any process for its construction.

Definition 3.5 ([25]) We define the weak variation $V(F, E)$, and the strong variation, $V_*(F, E)$, by

$$V(F, E) = \sup\left\{\sum_{i=1}^{n} |F(b_i) - F(a_i)|\right\}$$

and

$$V_*(F, E) = \sup\left\{\sum_{i=1}^{n} w(F, [a_i, b_i])\right\},$$

where the supremum is taken over all possible finite collections of nonoverlapping intervals that have end points in E.

1. We say that F is of bounded variation on E, (BV), if $V(F, E) < \infty$.
2. We say that F is of restricted bounded variation on E, (BV_*), if $V_*(F, E) < \infty$.
3. We say that F is absolutely continuous on E, (AC), if for each $\epsilon > 0$, there exists a $\delta > 0$ such that, for every collection $\{[a_i, b_i], 1 \leq i \leq n\}$, of nonoverlapping intervals with end points in E and $\sum_{i=1}^{n}(b_i - a_i) < \delta$, then $\sum_{i=1}^{n} |F(b_i) - F(a_i)| < \epsilon$.

4. We say that F is absolutely continuous on E in the restricted sense, $(AC)_*$, if for each $\epsilon > 0$, there exists a $\delta > 0$ such that, for every collection $\{[a_i, b_i], 1 \le i \le n\}$, of nonoverlapping intervals with end points in E and $\sum_{i=1}^{n}(b_i - a_i) < \delta$, then

$$\sum_{i=1}^{n} w(F, [a_i, b_i]) < \epsilon.$$

5. We say that F is generalized absolutely continuous on E, (ACG), if $F|_E$ is continuous and E is a countable union of sets $\{E_i\}$ such that F is (AC) on each E_i.
6. We say that F is generalized absolutely continuous in the restricted sense in E, $(ACG)_*$, if $F|_E$ is continuous and E is a countable union of sets $\{E_i\}$ such that F is $(AC)_*$ on each E_i.

The one thing that the Riemann approach in a first-year analysis course cannot do is allow an immediate entry into measure theory in the subsequent course. However, even here things are more interesting than this simple remark would suggest. Due to an idea of Henstock-Kurzweil, developed in particular by Thomson, there are associated with each Riemann approach a series of natural metric measures that on analysis are found to contain the basic information about the primitives more precisely than the classical concepts of ACG_*, etc. In particular, the concept of variation allows yet another way of defining the various Riemann integrals. These ideas are a little too refined for full mention here but are part of the excellent re-working of trigonometric integrals by Thomson.

Definition 3.6 Descriptive Definitions: Let E be a measurable subset of $\mathbb{R}$ and $\mu_\infty(E)$ denotes the Lebesgue measure of E. Let $c \in \mathbb{R}$.

1. We say that c is a point of density for E if

$$d_c E = \lim_{h \to 0^+} \frac{\mu_\infty(E \cap (c - h, c + h))}{2h} = 1.$$

2. We say that c is a point of dispersion for E if

$$d_c E = \lim_{h \to 0^+} \frac{\mu_\infty(E \cap (c - h, c + h))}{2h} = 0.$$

3. We say that a function $F : [a, b] \to \mathbb{R}$ is approximately continuous at $c \in E \subset [a, b]$, if c is a point of density for E and $F|_E$ is differentiable at c.

Theorem 3.1 ([25]) *Let F be a function defined on $[a, b]$ with $F(a) = 0$, then the following holds:*

1. *If F is (AC) on $[a, b]$, then F' exists a.e. and F' is Lebesgue integrable, then* $\int_a^x F'(y)d\mu_\infty(y) = F(x)$.
2. *If F is (ACG_*) on $[a, b]$, then F' exists a.e. and F' is Henstock-Kurzweil integrable, then* $\int_a^x F'(y)d\mu_\infty(y) = F(x)$.

Example 3.1 If F is any interval function and if $V_*F(E) = \inf_\delta \sup_P \sum |F(J_i)|$, where δ is a gauge and P is a partition in I that is anchored in the set E, then V_* is a regular Borel measure.

If V_* is AC, then F is the Henstock-Kurzweil integral of its derivative, which exists almost everywhere; further, if in addition, V_* is finite then F is the Lebesgue integral of this derivative.

Assume Ω be set of intervals, Σ are the σ-algebra of Ω. Throughout, the article $(\Omega, \Sigma, \mu_\infty)$ be finite measure space. Let [0, 1] be the unit interval of the real line with usual topology and μ_∞ be Lebesgue measure. In the whole article $\mathcal{X}$ is an arbitrary Banach space with dual $\mathcal{X}^*$. The closed unit ball of X is denoted by $B(\mathcal{X})$. If $Y \subset \mathcal{X}$, then $Y^\perp$ are annihilator of Y in $\mathcal{X}^*$. We denote $\sigma(\mathcal{X}, \mathcal{X}^*)$ are weak topology and the topology generated on $\mathcal{X}$ by a Banach space Y is $\sigma(\mathcal{X}, Y)$.

3.2 Vector Valued Henstock-Kurzweil Integrals

Definition 3.7 ([43, Definition 1]) A function $f : I = [a, b] \to \mathcal{X}$ is said to be Henstock-Kurzweil integrable on I if for every $\epsilon > 0$ there exists a gauge $\delta : I \to (0, +\infty)$, and $w \in \mathcal{X}$ such that for every δ-fine tagged partition $P = \{(t_i, I_i) : i = 1, 2, \ldots, p\}$ of I the following inequality

$$||\sum_{i=1}^{p} f(t_i)\mu_\infty(I_i) - w|| < \epsilon$$

holds. We set $w = HK \int_a^b f d\mu_\infty$.

If a subset $E \subset I$ is given, a function $f : I \to \mathcal{X}$ is integrable over the set E if the function $f.\chi_E : I \to \mathcal{X}$ is integrable.

Theorem 3.2 *Let $f : I \to \mathcal{X}$. If $f = 0$ a.e. in I, then f is Henstock-Kurzweil integrable on I and* $HK \int_a^b f d\mu_\infty = 0$.

Theorem 3.3 *A function $f : I \to \mathcal{X}$ is Henstock-Kurzweil integrable if and only if for every $\epsilon > 0$ there exists a gauge δ such that for every δ-fine partitions $P =$*

$\{(t_i, I_i) : i = 1, 2, \ldots, p\}$ *and* $Q = \{(s_j, J_j) : j = 1, 2, \ldots, r\}$ *of* I *the inequality* $\| \sum_{i=1}^{p} f(t_i)\mu_\infty(I_i) - \sum_{j=1}^{n} f(s_j)\mu_\infty(J_j)\| < \epsilon$ *holds.*

Theorem 3.4 *Assume* $f : I \to \mathcal{X}$ *is Henstock-Kurzweil integrable and* $J \subset I$ *be a compact interval. Then* f *is Henstock-Kurzweil integrable over* J.

Theorem 3.5 *Let* $f, g : I \to \mathcal{X}$ *be Henstock-Kurzweil integrable functions on* I. *Let* $c \in \mathbb{R}$, *then*

1. cf *is Henstock-Kurzweil integrable on* I.
2. $f + g$ *is Henstock-Kurzweil integrable on* I *and* $HK \int_a^b (f + g) d\mu_\infty = HK \int_a^b f d\mu_\infty + HK \int_a^b g d\mu_\infty$.

The previous theorem shows that Henstock-Kurzweil integrable functions hold linearity. We call $HK\ (I, \mathcal{X})$ the set of all Henstock-Kurzweil integrable functions $f : I \to \mathcal{X}$. The $HK(I, \mathcal{X})$ is endowed with the Alexiewicz norm

$$||h||_A = \sup_{a \le \alpha \le b} |HK \int_a^\alpha h(t) d\mu_\infty(t)|.$$

The completion of the $HK[a, b]$ is isomorphic to the space of distribution which are distributional derivatives of continuous function. The conjugate space $HK^*[a, b]$ is linearly isometric to the space $BV[a, b]$.

Definition 3.8 A function $f : [a, b] \to \mathcal{X}$ is said to be scalarly Henstock-Kurzweil integrable if for each $x^* \in \mathcal{X}^*$ the function $x^* f$ is Henstock-Kurzweil integrable. A scalarly Henstock-Kurzweil integrable function f is said to be Henstock-Kurzweil-Pettis integrable if for each $J \in [a, b]$ there exists $w_J \in \mathcal{X}$ such that

$$< x^*, w_J >= HK \int_J x^* f d\mu_\infty \ for\ every\ x^* \in \mathcal{X}^*.$$

We call w_J the Henstock-Kurzweil-Pettis integral of f over J and we write $w_J = (HKP) \int_J f d\mu_\infty$.

We call the additive interval function $F(J) = (HKP) \int_J f d\mu_\infty$ the HKP-primitive of f. Let $HKP([a, b], \mathcal{X})$ be the set of all $\mathcal{X}$-valued HKP-integrable function on $[a, b]$ (functions that are scalarly equivalent are identified). Henstock-Kurzweil-primitive (resp HKP-primitives) F of a function f is continuous (resp weakly continuous, i.e., $x^* F$ is continuous for every $x^* \in \mathcal{X}^*$). The Alexiewicz norm on $HKP([a, b], \mathcal{X})$ is

$$||f||_A = \sup_{a \le t \le b} \left\| (HKP) \int_a^b f d\mu_\infty \right\|.$$

If $|x^*f| \leq K_0$ a.e. for some constant K_0 and every $x^* \in \mathcal{X}^*$, then f is weakly bounded. If $f : [0, 1] \to \mathcal{X}$ is a scalarly bounded function, then the image measure λ given by

$$\lambda(E) = \mu_\infty(f^{-1}(E))$$

is a weak Baire measure on $\mathcal{X}$.

Definition 3.9 Given a weak Baire measure μ_∞ on a Banach space $\mathcal{X}$, we say μ_∞ is scalarly concentrated on a subspace $\mathbb{G}$ of $\mathcal{X}$ if $x^*|_{\mathbb{G}} = 0$ implies $x^* = 0$ μ_∞-a.e.

Recalling a topological space $\mathcal{X}$ is measure compact if every nonzero-σ-additive Baire measure on $\mathcal{X}$ has non-empty support. Measure-compact implies real compact but not conversely. It is well known that every compact measure is perfect. For a measure to be perfect, it is necessary and sufficient that its restriction to any σ-subalgebra with a countable number of generators be compact.

A Banach space $\mathcal{X}$ is said to have Pettis integral property (PIP) if for every $\mathcal{X}$-valued weakly bounded function f defined on an arbitrary finite measure space is Pettis integrable. We say that $\mathcal{X}$ has Mazur's property if each $weak^*$-continuous functional on X^* is in $\mathcal{X}$.

Definition 3.10 A scalarly measurable function $f : [a, b] \to \mathcal{X}$ is determined by a space $Y \subseteq \mathcal{X}$ if for every $x^* \in Y^\perp$ the equality $x^* f = 0$ holds μ_∞-a.e.

Theorem 3.6 *If $f : [a, b] \to \mathcal{X}$ is Henstock-Kurzweil-Pettis integrable, then there exists a sequence (f_n) of X-valued step function such that for all $x^* \in \mathcal{X}^*$, $x^* f = \lim_n x^* f_n$ a.e. if and only if f is determined by a separable subspace of $\mathcal{X}$.*

References

1. Alexiewicz, A.: Linear functionals on Denjoy-integrable functions. Collectanea Mathematica **1**, 289–293 (1948)
2. Boccuto, A., Minotti, A.M., Sambucini, A.R.: Set-valued Kurzweil-Henstock integral in Riesz spaces. Panamerican Math. J. **23**(1), 57–74 (2013)
3. Boccuto, A., Candeloro, D., Sambucini, A.R.: Henstock multivalued integrability in Banach lattices with respect to pointwise non atomic measures. Rendiconti Lincei-Matematica e Applicazioni **26**(4), 363–383 (2015)
4. Bongiorno, B., Di. Piazza, L., Musiał', K.: Approximation of Banach space valued Non-absolutely integrable functions by step functions. Glasgow Math. J. **50**, 583–593 (2008)
5. Candeloro, D., Di Piazza, L., Musiał, K., Sambucini, A.R.: Some new results on integration for multifunctions. Ricerche di Matematica **67**, 361–372 (2018)
6. Candeloro, D., Di Piazza, L., Musiał, K., Sambucini, A.R.: Gauge integrals and selections of weakly compact valued multifunctions. J. Math. Anal. Appl. **441**(1), 293–308 (2016)
7. Candeloro, D., Di Piazza, L., Musiał, K., Sambucini, A.R.: Some new results on integration for multifunction. Ricerche di Matematica **67**(2), 361–372 (2018)
8. Candeloro, D., Sambucini, A.R.: Comparison between some norm and order gauge integrals in Banach lattices. Panamerican Math. J. **25**(3), 1–16 (2015)

9. Candeloro, D., Sambucini, A.R.: Order-type Henstock and McShane integrals in Banach lattice setting. In: Proceedings of the SISY 2014 - IEEE 12th International Symposium on Intelligent Systems and Informatics, pp. 55–59 (2014)
10. Candeloro, D., Di Piazza, L., Musial, K., Sambucini, A.R.: Gauge integrals and selections of weakly compact valued multifunctions. J. Math. Anal. Appl. **441**, 293–308 (2016)
11. Candeloro, D., Di Piazza, L., Musial, K., Sambucini, A.R.: Some new results on integration for multifunctions. Ricerche di Matematica **67**, 361–372 (2018)
12. Candeloro, D., Di Piazza, L., Musial, K., Sambucini, A.R.: Integration of multifunctions with closed convex valued in arbitrary Banach spaces. J. Convex Anal. **27**(4), 1233–1246 (2020)
13. Cao, S.: The Henstock integral for Banach-valued functions. Southeast Asian Bull. Math. **16**(1), 35–40 (1992)
14. Davies, R.O., Schuss, Z.: A proof that Henstock's integral includes Lebesgue's. J. Lond. Math. Soc. **2**, 561–562 (1970)
15. Di Piazza, L.: Kurzweil-Henstock type integration on Banach spaces. Real Anal. Exchange **29**(2), 543–555 (2003/2004)
16. Di Piazza, L., Musiał, K.: Set valued Kurzweil-Henstock-Pettis integral. Set-Valued Anal. **13**, 167–169 (2005)
17. Di Piazza, L., Musiał, K.: Characterizations of Kurzweil-Henstock-Pettis integrable functions. Studia Mathematica **176**(2), 159–176 (2006)
18. Di Piazza, L., Marraffa, V., Musial, K.: Variational Henstock integrability of Banach space valued functions. Mathematica Bohemica **141**, 287–296 (2016)
19. Di Piazza, L., Musial, K.: Set valued Kurzweil-Henstock Pettis integral. Set-Valued Anal. **13**, 167–169 (2005)
20. Federson, M.: Some Peculiarities of the HK-integrals of Banach space-valued functions. Real Anal. Exchange **29**(1), 439–460 (2003/2004)
21. Forkert, D.L.: The Banach space-valued integrals of Riemann, Mcshane, Henstock-Kurzweil and Bochner, Bachelor Thesis, Vienna University of Technology (2012)
22. Fong, C.K.: A continuous version of Orlicz-Pettis theorem via vector-valued Henstock-Kurzweil integrals. Can. Math. Bull. **24**(2), 169–176 (1981)
23. Fremlin, D.H., Mendoza, J.: On the integration of vector-valued functions. Illinois J. Math. **38**(1), 127–147 (1994)
24. Gamez, J.L., Mendoza, J.: Denjoy-Dunford and Denjoy-Pettis integral. Studia Mathematica **130**(2), 115–133 (1998)
25. Gill, T.L., Zachary, W.W.: Functional Analysis and Feynman Operator Calculus. Springer, Cham (2016)
26. Gordon, R.A.: The Denjoy extension of the Bochner, Pettis, and Dunford integrals. Studia Mathematica **92**(1), 73–91 (1989)
27. Gordon, R.: The Integrals of Lebesgue, Denjoy, Perron, and Henstock. American Mathematical Society, Providence (1994)
28. Guoju, Y., Tianguing, A.N.: On Henstock-Dunford and Henstock-Pettis integral. Int. J. Math. Math. Sci. **25**(7), 467–478 (2001)
29. Guoju, Y., Schwabik, S.: The Mcshane and weak Mcshane integrals of Banach space-valued functions defined on $\mathbb{R}^m$. Math. Notes Miskolc **2**(2), 127–136 (2001)
30. Grabiner, J.V.: Who gave you the Epsilon Cauchy and the origins of rigorous calculus. Am. Math. Monthly **90**(3), 185–194 (1983)
31. Henstock, R.: The General Theory of Integration. Oxford Mathematical Monographs. Clarendon Press, Oxford (1991)
32. Henstock, R.: The equivalence of generalized forms of the Ward, variational, Denjoy-Stieltjes, and Perron-Stieltjes integrals. Proc. Lond. Math. Soc. **3**(10), 281–303 (1960)
33. Henstock, R.: Lectures on the Theory of Integration. World Scientific, Singapore (1988)
34. Henstock, R.: A Riemann-type integral of Lebesgue power. Can. J. Math. **20**, 79–87 (1968)
35. Henstock, R.: Definitions of Riemann type of the variational integrals. Proc. Lond. Math. Soc. **3**(11), 402–418 (1961)

36. Hoffmann, H.: Descriptive Characterisation of the Variational Henstock-Kurzweil-Stieltjes Integral and Applications. Karlsruhe Institute of Technology, Faculty of Mathematics, Institute for Analysis (2014)
37. Kurzweil, J.: Generalized ordinary differential equations and continuous dependence on a parameter. Czech. Math. J. **7**(82), 418–446 (1957)
38. Lebesgue, H.: Intégrale, longueur, aire. Annali di Matematica Pura ed Applicata **7**(1), 231–359 (1902)
39. Lee, P.Y., Výborný, R.: The Integral: An Easy Approach After Kurzweil and Henstock. Cambridge University Press, Cambridge (2000)
40. Lindenstrauss, J., Tzafriri, L.: Classical Banach Spaces II, Function Spaces. A Series of Modern Surveys in Mathematics. Springer, Berlin-Heidelberg-New York (1979)
41. Precupanu, A.M., Croitoru, A.: A Gould type integral with respect to a multimeasure II. Analele Universitatii Alexandru Ioan Cuza Iaşi XLIX 183–207 (2003)
42. Schwabik, S., Guoju, Y.: Topics in Banach Spaces Integration, Series in Real Analysis, vol 10. World Scientific, Singapore (2005)
43. Swartz, C.: Introduction to Gauge Integral. World Scientific, Singapore (2001)
44. Yeong, L.T.: Henstock-Kurzweil Integration on Euclidean Spaces. Series in Real Analysis, vol. 12. World Scientific Publishing, Singapore (2011)

Chapter 4
Kluvánek-Lewis-Henstock Integrals

4.1 Introduction

In 1955, Bartle, Dunford, and Schwartz developed a theory of integration for scalar functions with respect to a σ-additive Banach-space-valued vector measure μ_∞ defined on a σ-algebra of sets and used it to give an integral representation for weakly compact operators $u : C(S) \to \mathcal{X}$, where S is a compact Hausdorff space and $\mathcal{X}$ is a Banach space (see [28, 38]). About 15 years later, Lewis studied a Pettis-type weak integral of scalar functions with respect to a σ-additive vector measure μ_∞ having range in a locally convex Hausdorff $\mathcal{X}$ (see [26, 27]. This type of definition has also been considered by Kluvánek in [24]. In honor of these mathematicians in the literature, mathematicians call the integral introduced in [26, 27] as well as its variant is called Kluvánek-Lewis integral or briefly, the (KL)-integral. [26, Theorem 2.4] asserts that the (KL)-integral is the same as the (BDS)-integral. Also see [23, 24, 41] for details about (KL)-integral. In the 1960s, both Henstock and Kurzweil independently developed the so-called Henstock-Kurzweil (HK) integral; an integral able to integrate a much larger class of functions than the Riemann integral and the Lebesgue integral, compatible with the Fundamental Theorem of Calculus. Recalling the Henstock-Kurzweil integral provides a tool for integrating highly oscillatory functions which occur in nonlinear analysis and in quantum theory. It is also easy to understand because its definition requires no measure theory. Moreover, all Bochner integrable (in real-valued case Lebesgue integrable) functions are Henstock-Kurzweil (shortly HK) integrable, but not conversely. Now we recall the Henstock-Kurzweil (in short, (HK))-integral for real-valued functions, defined on abstract sets, with respect to (possibly infinite) nonnegative measures. For a related literature, see [1, 3, 5, 6, 8–10, 12, 13, 17–19, 25, 33, 37]. Candeloro et al. [6, 7] in recent studies about integration theory have said that multifunctions have played an important role in applications to several branches of science, like control theory, differential inclusions, game theory,

B. Hazarika, H. Kalita, *Theory of Henstock–Orlicz Spaces*, Mathematical Marvels: Texts and Monographs in the Spirit of CR Rao,
https://doi.org/10.1007/978-981-96-9548-5_4

aggregation functions, economics, problems of finding equilibria, and optimization. Since neither the Riemann Integral, nor the Lebesgue integral is completely satisfactory concerning the problem of the existence of primitives, different types of integrals extending the previous ones have been introduced and investigated, like Henstock-Kurzweil, McShane, and Pettis integrals. These topics have many connections with measures taking values in abstract spaces, and in particular, the extension of the concept of integrability to set-valued functions can be used in order to obtain a larger number of selections for multifunctions, through their estimates and properties, in several applications, see [5, 6, 8, 9]. In recent times, Boccuto et al. [2] introduced the non-absolute version of (KL)-integral for their purpose. N.S. papageorgious introduced the concept of the integral of set-valued measure by the bilinear integral of Dinculeanu, see [14, 31]. Kandilakis introduced an integral of bounded real-valued functions with respect to a set-valued measure using the set of Kluvánek-Knowles-type integrals, see [22]. Wu et al. introduced the set-valued Bartle integral which is the set of Bartle-Dunford-Schwartz-type integrals. The set-valued Bartle integral has also been deeply considered by Zhang et al. Precupanu and Croitoru introduced a multivalued Gould-type integral for a real-valued function with respect to a set-valued measure, see [34, 35, 40, 42]. Cai Li Zhou and Fu Gui SHI introduce and study a new set-valued integral of scalar-valued functions with respect to a set-valued measure, see [41]. Unlike various ways of defining such an integral (e.g., Papageorgiou way, Kandilakis way, and Wu, Zhang, and Wang way), their integral does not depend on measure selections of the given set-valued measure and is a Pettis-type set-valued integral of scalar-valued functions with respect to a set-valued measure. The result is an extension of the Kluvánek-Lewis-type integral, which is a Pettis-type weak integral of scalar-valued functions with respect to a vector-valued measure. In this chapter we study some properties and convergence theorem of "Kluvánek-Lewis-Henstock μ_∞-integrability" for "μ_∞-measurable functions." We introduce "Kluvánek-Lewis-Henstock integrable" of scalar-valued functions with respect to a set-valued measure in a Banach space. Finally we introduce Lebesgue-type dominated convergence theorem for the set-valued "Kluvánek-Lewis-Henstock integral."

4.2 Basic Definitions and Results

Let $T \neq \emptyset$ be an abstract set, $P[T]$ be the class of all subsets of T, $\Sigma \subset P[T]$ be a σ-algebra , $\mathcal{X}$ be a Banach space, and $\mathcal{X}^*$ be its topological dual. For each $A \in \Sigma$, let us denote by χ_A the *characteristic function* of A, defined by:

$$\chi_A(t) = \begin{cases} 1 \text{ if } t \in A, \\ 0 \text{ if } t \in T \setminus A. \end{cases}$$

Assume T is a compact topological space and Σ is the σ-algebra of all Borel subsets of T. Let $\nu : \Sigma \to \mathbb{R} \cup \{+\infty\}$ be a σ-additive nonnegative measure. A vector measure is a σ-additive set function $\mu_\infty : \Sigma \to \mathcal{X}$. By the Orlicz-Pettis theorem , see also [11, Corollary 1.4], the σ-additivity of μ_∞ is equivalent to the σ-additivity of the scalar-valued set function $x^*\mu_\infty : A \mapsto x^*(\mu_\infty(A))$ on Σ for every $x^* \in \mathcal{X}^*$. For a literature on vector measures, see also [7, 11, 23, 27, 30, 32]. To make our presentation reasonably self-contained we recall a few definitions and results in this section that we will use in our main section.

The variation $|\mu_\infty|$ of μ_∞ is defined by setting

$$|\mu_\infty|(A) = \sup \left\{ \sum_{i=1}^{r} \|\mu_\infty(A_i)\| : A_i \in \Sigma, i = 1, 2, \ldots, r; A_i \cap A_j = \emptyset \right\}$$

for $i \neq j$; $\bigcup_{i=1}^{r} A_i \subset A$. We define the semivariation $\|\mu_\infty\|$ of μ_∞ by

$$\|\mu_\infty\|(A) = \sup_{x^* \in \mathcal{X}^*, \|x^*\| \leq 1} |x^*\mu_\infty|(A). \tag{4.1}$$

Remark 4.1 Observe that $\|\mu_\infty\|(A) < +\infty$ for all $A \in \Sigma$ (see also [11, Corollary 1.19], [30, §1]).

The *completion* of Σ with respect to $\|\mu_\infty\|$ is defined by

$$\widetilde{\Sigma} = \{A = B \cup N : B \in \Sigma, N \subset M \in \Sigma \text{ with } \|\mu_\infty\|(M) = 0\}. \tag{4.2}$$

A function $f : T \to \mathbb{R}$ is said to be μ_∞-measurable if

$$f^{-1}(B) \cap \{t \in T : f(t) \neq 0\} \in \widetilde{\Sigma}$$

for each Borel subset $\mathfrak{B} \subset \mathbb{R}$.

From (4.1) and (4.2) it follows that every μ_∞-measurable real-valued function is also $x^*\mu_\infty$-measurable for every $x^* \in \mathcal{X}^*$. Moreover, it is readily seen that every Σ-measurable real-valued function is also μ_∞-measurable. The complement of $\sigma(P)$ is $\overline{\sigma(P)}$ with respect to the semivariation $||\mu_\infty||$. We recall the definition of $\overline{\sigma(P)}$ as follows:

Definition 4.1 ([15, Definition 11]) $\overline{\sigma(P)} = \{E = F \cup N : F \in \sigma(P), N \subset M \in \sigma(P)$ *with* $||\mu_\infty|| = 0\}$.

Recalling that $||\mu_\infty||$ is a σ-subadditive submeasure on $\sigma(P)$, then $\overline{\sigma(P)}$ is a σ-ring and that $\overline{\sigma(P)}$ contains $\sigma(P)$. We say that μ_∞ is Σ-separable (or separable) if there is a countable family $\mathbb{B} = (B_k)_k$ in Σ such that for each $A \in \Sigma$ and $\varepsilon > 0$, there is $k_0 \in \mathbb{N}$ such that

$$\|\mu_\infty\|(A \Delta B_{k_0}) = \sup_{x^* \in X^*, \|x^*\| \le 1} [\,|x^* \mu_\infty|(A \Delta B_{k_0})] \le \varepsilon, \tag{4.3}$$

see also [36]. Such a family $\mathbb{B}$ is said to be μ_∞-dense .

Observe that μ_∞ is Σ-separable if and only if Σ is μ_∞-*essentially countably generated*, namely there is a countably generated σ-algebra $\Sigma_0 \subset \Sigma$ such that for each $A \in \Sigma$ there is $B \in \Sigma_0$ with $\mu_\infty(A \Delta B) = 0$. The separability of μ_∞ is satisfied, for instance, when T is a separable metrizable space, Σ is the Borel σ-algebra of the Borel subsets of T, and μ_∞ is a Radon measure (see also [4, Theorem 4.13], [16, Theorem 1.0], [23, §1.3 and §2.6], [36, Propositions 1A and 3]). A decomposition of a set $A \in \Sigma$ is a finite collection $\{(A_1, \xi_1), (A_2, \xi_2), \dots, (A_N, \xi_N)\}$ such that $A_j \in \Sigma$ and $\xi_j \in A_j$ for every $j \in \{1, 2, \dots, N\}$, and $\nu(A_i \cap A_j) = 0$ whenever $i \neq j$. A decomposition of subsets of $A \in \Sigma$ is called a partition of A when $\bigcup_{j=1}^{N} A_j = A$. A gauge on a set $A \in \Sigma$ is a map δ assigning to each point $x \in A$ a neighborhood $\delta(x)$ of x. If $\mathcal{D} = \{(A_1, \xi_1), (A_2, \xi_2), \dots, (A_N, \xi_N)\}$ is a decomposition of A and δ is a gauge on A, then we say that $\mathcal{D}$ is δ-fine if $A_j \subset \delta(\xi_j)$ for any $j \in \{1, 2, \dots, N\}$.

Any gauge in $[a, +\infty]$ has the form $\delta(x) = (x - d(x), x + d(x))$, if $x \in [a, +\infty] \cap \mathbb{R}$, and $\delta(+\infty) = (b, +\infty] = (b, +\infty) \cup \{+\infty\}$, where d denotes a positive real-valued function defined on $[a, +\infty)$. Now we define the Riemann sums by $S(f, \mathcal{D}) = \sum_{j=1}^{N} f(\xi_j)\nu(A_j)$ if the sum exists in $\mathbb{R}$, with the convention $0 \cdot (+\infty) = 0$. Note that for any gauge δ there exists at least one δ-fine partition $\mathcal{D}$ such that $S(f, \mathcal{D})$ is well-defined. A *vector measure* is a σ-additive set function $\mu_\infty : \Sigma \to \mathcal{X}$. By the Orlicz-Pettis theorem (see also [11, Corollary 1.4]), the σ-additivity of μ_∞ is equivalent to the σ-additivity of the scalar-valued set function $x^*\mu_\infty : A \mapsto x^*(\mu_\infty(A))$ on Σ for every $x^* \in \mathcal{X}^*$. For a literature on vector measures, see also [7, 11, 23, 27, 30, 32].

Definition 4.2 ([2, 20]) A function $f : T \to \mathbb{R}$ is said to be Henstock-Kurzweil integrable (briefly HK-integrable) on a set $A \in \Sigma$ if there is an element $I_A \in \mathbb{R}$ such that for every $\varepsilon > 0$ there is a gauge δ on A with $|S(f, \mathcal{D}) - I_A| \le \varepsilon$ whenever $\mathcal{D}$ is a δ-fine partition of A such that $S(f, \mathcal{D})$ exists in $\mathbb{R}$, and we write

$$HK \int_A f \, d\nu = I_A.$$

Definition 4.3 ([2]) We say that a Σ-measurable function $f : T \to \mathbb{R}$ is Kluvánek-Lewis-Henstock-Kurzweil μ_∞-integrable , shortly (HKL) μ_∞-integrable if the following properties hold:

$$f \text{ is } |x^*\mu_\infty|\text{-HK-integrable for each } x^* \in X^*, \tag{4.4}$$

and for every $A \in \Sigma$ there is $x_A^{(HK)} \in X$ with

$$x^*(x_A^{(HK)}) = HK \int_A f\, d|x^*\mu_\infty| \text{) for all } x^* \in X^*, \tag{4.5}$$

where the symbol (HK) in (4.5) denotes the usual Henstock-Kurzweil integral of a real-valued function with respect to a (extended) real-valued measure.

A Σ-measurable function $f : T \to \mathbb{R}$ is said to be *weakly (HKL)* μ_∞*-integrable* weakly (HKL) μ_∞-integrable if it satisfies only condition (4.4). If $f : T \to \mathbb{R}$ is a (HKL)-integrable function, then the Alexiewicz norm of f is defined by

$$||f||_{HKL} = \sup_{x^* \in \mathcal{X}^*, \|x^*\| \le 1} \left(\sup_{A \in \Sigma} \left| HK \int_A f(t)\, d|x^*\mu_\infty| \right| \right)$$

(see also [39]). Observe that, by arguing analogously as in [19, Theorem 9.5] and [29, Example 3.1.1], for each $x^* \in \mathcal{X}^*$ we get that $f = 0$ $|x^*\mu_\infty|$ almost everywhere if and only if $HK \int_A f(t)\, d|x^*\mu_\infty| = 0$ for every $A \in \Sigma$. Thus, it is not difficult to see that $\|\cdot\|_{HKL}$ is a norm. In general, the space of the real-valued Henstock-Kurzweil integrable functions endowed with the Alexiewicz norm is not complete.

Theorem 4.1 ([11, Corollary 1.4]) *Let $\mathcal{X}$ be a Banach space. Then a series $\sum_{n=1}^{\infty} x_n$ in X is unconditionally norm convergent whenever it is weakly subseries convergent.*

Proposition 4.1 ([32, Proposition 1.1.1]) *Let $\mathcal{R}$ be a ring of sets in T and let μ_∞ : $\mathcal{R} \to \mathcal{X}$ be additive. Then*

$$||\mu_\infty||(E) = \sup_{|x^*| \le 1} \nu(x^*\mu_\infty)(E) \ for\ E \in \sigma(\mathcal{R}).$$

Proposition 4.2 ([14, Proposition 5]) *Let P be a δ-ring of subsets of T and let ν be a (σ-additive) scalar measure on P. If $f : T \to \mathbb{K}$ is ν-integrable, let $\nu(E) = \int_E f d\nu$, $E \in \sigma(P)$, then ν is σ-additive on $\sigma(P)$ where $\mathbb{K}$ is $\mathbb{C}$ or $\mathbb{R}$.*

Proposition 4.3 ([27, Theorem 1.3]) *Let Σ be a σ-ring of sets. If $\nu : \Sigma \to \mathcal{X}$ is σ-additive and if (E_n) is a convergent sequence in Σ with its limit E, then $\lim_n ||\nu||(E_n||(E_n) = ||\nu||(E)$.*

Theorem 4.2 (Pettis) *Let Σ be a σ-ring of subsets of T. Let $\lambda : \Sigma \to [0, \infty]$ be a σ-subadditive submeasure and let $\nu : \Sigma \to \mathcal{X}$ be a σ-additive, then $\nu << \lambda$ if and only if ν is λ-continuous.*

Proposition 4.4 ([32, Proposition 1.1.12]) *A continuous submeasure λ defined on a σ-ring $\mathcal{R}$ is a σ-subadditive.*

We denote by $\sigma(., A)$ the support function of the set $A \subset \mathcal{X}$ defined by

$$\sigma(x^*, A) = \sup_{x \in A} x^*(x), \;\; x^* \in \mathcal{X}^*$$

Let $P_0(X) = \{A \subset \mathcal{X} : \text{A is non empty subset of } \mathcal{X}\}$. Let $A, B \in P_f(\mathcal{X})$, the Hausdorff metric H of A & B is defined by

$$H(A, B) = \max\{\sup_{a \in A} d(a, B), \;\; \sup_{b \in B} d(b, A)\},$$

where $d(a, B) = \inf_{b \in B}\{\sup d(a, B), \;\; \sup_{b \in B} d(b, A)\}$ and $d(a, B) = \inf_{b \in B} ||a - b||$.

Definition 4.4 ([41, Definition 1]) Let (T, Σ) be a measurable space. A set-valued set function $\mathcal{M} : \Sigma \to P_0(\mathcal{X})$ is said to be a set-valued measure if it satisfies the following two conditions:

1. $\mathcal{M}(\emptyset) = \{0\}$
2. If $A_1, A_2, \ldots$ are in Σ with $A_i \cap A_j = \emptyset$ for $i \neq j$, then $\mathcal{M}(\sum_{n=1}^{\infty} A_n) = \sum_{n=1}^{\infty} \mathcal{M}(A_n)$, where

$$\sum_{n=1}^{\infty} \mathcal{M}(A_n) = \{x \in X : x = \sum_{n=1}^{\infty} x_n \; (unc.conv) \; , \; x_n \in \mathcal{M}(A_n), \; n \geq 1\}.$$

In particular $\mathcal{M} : \Sigma \to P_{wkc}(X)$ is a set-valued measure iff $\sigma(x^*, \mathcal{M}(.))$ is a scalar-valued measure for all $x^* \in X^*$. A vector-valued measure $\mu_\infty : \Sigma \to \mathcal{X}$ satisfying $\mu_\infty(A) \in \mathcal{M}(A), \forall \; A \in \Sigma$ is said to be a selection of the set-valued measure $\mathcal{M}$. We denote the set of all selections of $\mathcal{M}$ by $\mathcal{S}_{\mathcal{M}}$.

Recalling that the support function satisfies the property

$$\sigma(x^*, A + B) = \sigma(x^*, A) + \sigma(x^*, B)$$
$$\sigma(x^*, \lambda A) = \lambda(x^*, A)$$

for all $A, B \in P_0(X), \; x^* \in \mathcal{X}^*$ & $\lambda \geq 0$.

In particular

$$H(A, B) = \sup_{||x^*|| \leq 1} |\sigma(x^*, A) - \sigma(x^*, B)|$$

whenever A, B are two convex sets.

4.3 Basic Properties of Kluvánek-Lewis-Henstock Integrable μ_∞-Integral

In this section we will discuss basic properties of (HKL) μ_∞-integrable function which are μ_∞-measurable.

Definition 4.5 ([21]) If f is *μ_∞-measurable*, then there exists $M \in \sigma(P)$ with $||\mu_\infty||(M) = 0$ such that $f_{\chi_{T \setminus M}}$ is *(HKL) μ_∞-integrable* in T, then we say that f is *(HKL) μ_∞-integrable* in T, and in that case we define

$$(HKL) \int_E f d|x^* \mu_\infty| = (HKL) \int_E f_{\chi_{T \setminus M}} d|x^* \mu_\infty| \text{ for } E \in \sigma(P) \tag{4.6}$$

and

$$(HKL) \int_T f d|x^* \mu_\infty| = (HKL) \int_T f \chi_{T \setminus M} d|x^* \mu_\infty|. \tag{4.7}$$

Let f be μ_∞-measurable and let it be *(HKL) μ_∞-integrable* in T with M as in Definition 4.5. If $M_1 \in \sigma(P)$ with $||\mu_\infty||(M_1) = 0$ is such that $f_{\chi_{T \setminus M_1}}$ is *$\sigma(P)$-measurable* , then it can easily be shown that $f_{\chi_{T \setminus M_1}}$ is also *(HKL)μ_∞-integrable* in T and $(HKL) \int_E f_{\chi_{T \setminus M}} d|x^* \mu_\infty|$ for $E \in \sigma(P) \cup \{T\}$. Consequently $(HKL) \int_E f d|x^* \mu_\infty|$ is well defined and does not depend on the particular choice of the $|x^* \mu_\infty|$-null set M in Definition 4.5 of the above definition. From now onward we denote the notation $|x^* \mu_\infty|$ as m. $\mathcal{H}(\mu_\infty)$ denotes the class of all μ_∞-measurable scalar functions on T which are (HKL) μ_∞-integrable in T. It is observed that if $A, B \in \Sigma$, $B \subset A$ and $f : T \to \mathbb{R}$ is (HKL)-integrable on A, then f is also (HKL)-integrable on B and on $A \setminus B$, and

$$(HKL) \int_A f(t)\, dm = (HKL) \int_B f(t)\, dm + (HKL) \int_{A \setminus B} f(t)\, dm$$

Theorem 4.3 *Let $f : T \to K$ or $[-\infty, \infty]$ and let s be a P-*simple function *. Then the following hold:*

1. *If $s = \sum_1^r \alpha_i \chi_{E_i}$ is P-*simple *with $(\alpha_i)_1^r \subset K$, $(E_i)_1^r \subset P$ and $E_i \cap E_j = \emptyset$ for $i \neq j$, then s is μ_∞-integrable in T and $(HKL) \int_E s d\mu_\infty = \sum_1^r \alpha_i \mu_\infty(E_i \cap E)$*

for $E \in \sigma(P)$. Consequently

$$||\mu_\infty||(E) = \sup\left\{|(HKL)\int_E sdm| \ : s \ P\text{-}simple,\right.$$
$$\left.|s(t)| \leq \chi_E(t), \ t \in T\right\}.$$

2. *If f is (HKL)μ_∞-integrable in T, then $\nu(.) : \sigma(P) \to \mathcal{X}$ given by*

$$\nu(.) = (HKL)\int_{(.)} fdm \text{ is } \sigma\text{-additive in norm.}$$

3. *If $\nu(.)$ as in (2), then*
   (a) $||\nu||(E) = \sup_{|x^*|\leq 1} \int_E |f|d\nu(m), \ E \in \sigma(P)$
   (b) $\lim_{||\mu_\infty||\to 0} \nu(E) = \lim_{||\mu_\infty||(E)\to 0} ||\nu||(E) = 0, \ E \in \sigma(P).$
4. *$\mathcal{H}(\mu_\infty)$ is a vector space over K with respect to pointwise addition and scalar multiplication. Moreover, for $E \in \sigma(P)$ fixed, the mapping $f \to (HKL)\int_E fdm$ is linear on $\mathcal{H}(\mu_\infty)$. Consequently $\int_E sdm = \sum_{i=1}^{r} \alpha_i m(E_i \cap E)$ for $E \in \sigma(P)$ and for $s = \sum_1^r \alpha_i \chi_{E_i}$, with $(E_i)_1^r \subset P$ not necessarily mutually disjoint.*

Proof Without loss of generality we shall assume that all the m-measurable functions considered here are further $\sigma(P)$-measurable. (1) is very obvious. Using the Theorem 4.1 we can find the proof of (2).

For $3(a)$: Using (2) and Proposition 4.1 and the Proposition 4.2.

For $3(b)$ If $||\mu_\infty||(E) = 0, \ E \in \sigma(P)$, then $\nu(x^*\mu_\infty)(E) = 0$ for each $x^* \in \mathcal{X}^*$. Now by $3(a)$ $||\nu||(E) = 0$. As $||\mu_\infty||$ and $||\nu||$ are nonnegative, monotone, and σ-subadditive on the σ-ring $\sigma(P)$, then f is $(HKL)\mu_\infty$-integrable in T implies $\nu(.) = (HKL)\int_{(.)} fdm$ is σ-additive in norm. Also by Proposition(4.3) $||\nu||$ is continuous on $\sigma(P)$. From the fact $|\nu(E)| \leq ||\nu||(E)$, $E \in \sigma(P)$ finally gives

$$\lim_{||\mu_\infty||\to 0} \nu(E) = \lim_{||\mu_\infty||(E)\to 0} ||\nu||(E) = 0 \, , E \in \sigma(P).$$

Proof of (4) is very obvious. We leave for the readers. □

Theorem 4.4 *For a set $A \in T$, let $P \cap A = \{B \cap A : \ B \in P\}$. If f is a μ_∞-measurable function on T and if f is μ_∞-essentially bounded in a set $A \in P$, then f is (HKL)-μ_∞-integrable in each $B \in P \cap A$ and*

$$|(HKL)\int_B fd\mu_\infty| \leq (ess\sup_{t\in A} |f(t)|)||\mu_\infty||(B),$$

where $B \in \sigma(P) \cap A = P \cap A$.

Proof Let $A \in P$. Let $\Re = \operatorname{ess\,sup}_{t \in A} |f(t)|$. Then there exists $M_1 \in \sigma(P) \cap A = P \cap A$ with $||\mu_\infty||(M_1) = 0$ such that $\sup_{t \in A \setminus M_1} |f(t)| = \Re$. Let $\Sigma_A = P \cap A$. Then Σ_A is σ-algebra of subset of A. Also $f_{\chi_{A\setminus M_1}}$ is Σ_A-measurable and bounded, then there exists a sequence $(s_n)_{n=1}^{\infty}$ of Σ_A-simple function such that $s_n \to f_{\chi_{A\setminus M_1}}$ uniformly in A with $|s_n| \to f$ in A. Let $B \in \Sigma_A$. Then for $\epsilon > 0$ there exists n_0 such that $||s_n - s_m||_A||\mu_\infty||(A) < \epsilon$ for $n, m > n_0$. Consequently,

$$
\begin{aligned}
|HK \int_B s_n d\mu_\infty - HK \int_B s_m d\mu_\infty| &\leq ||s_n - s_m||_B||\mu_\infty||(B) \\
&\leq ||s_n - s_m||_A||\mu_\infty||(A) \\
&< \epsilon \text{ for } n, m \geq n_0.
\end{aligned}
$$

So, for each $B \in \Sigma_A$ there exists a vector $x_B^{(HK)} \in \mathcal{X}$ such that $\lim_{n\to\infty} (HK) \int_B s_n d\mu_\infty = x_B^{(HK)}$. Now by Lebesgue Dominated Convergence Theorem on σ-algebra Σ_A for scalar measure, we have

$$
x^*(x_B^{(HK)} = \lim_{n\to\infty} (HK) \int_B s_n d(x^* \mu_\infty)
$$

for each $x^* \in \mathcal{X}^*$ and for each $B \in \Sigma_A$, thus f is $(HKL) - \mu_\infty$integrable in A and

$$
(HKL) \int_B f d\mu_\infty = x_B^{(HK)}
$$

for each $B \in \Sigma_A$. So,

$$
\begin{aligned}
|(HKL) \int_B f d\mu_\infty| &= |x_B^{(HK)}| \\
&= |\lim_{n\to\infty} HK \int_B s_n d\mu_\infty| \\
&\leq \Re||\mu_\infty||(B)
\end{aligned}
$$

for $B \in \Sigma_A$. This completes proof. □

Theorem 4.5 *If* ϕ *is a* μ_∞*-essentially bounded* μ_∞*-measurable scalar function on* T *and if* $f \in \mathcal{H}(\mu_\infty)$*, then* $\phi.f$ *is (HKL)* μ_∞*-integrable in* T *and*

$$
(HKL) \int_E f d(|x^* \mu_\infty|) = (HKL) \int_T \chi_E f d(|x^* \mu_\infty|).
$$

Proof Let $\operatorname{ess\,sup}_{t\in T}|\phi(t)| = \mathfrak{K}$ and $\mu_\infty(.) = (HKL)\int_{(.)} f|x^*\mu_\infty|$ on $\sigma(P)$. Then μ_∞ is σ-additive on $\sigma(P)$ and $\sigma(P)$-measurable function ϕ is also μ_∞-essential bounded, hence ϕ is (HKL) μ_∞-integrable in T. Let $(HKL)\int_E \phi d\mu_\infty = x_E^{(HK)}$ for $E \in \sigma(P)$. Now

$$\begin{aligned} x^*\left(x_E^{(HK)}\right) &= HK\int_E \phi d(|x^*\mu_\infty|) \\ &= HK\int_E \phi f d(|x^*\mu_\infty|) \end{aligned}$$

for $E \in \sigma(P)$ and $x^* \in \mathcal{X}^*$. Using [32, Proposition 1.1.18], Lebesgue Dominated Convergence Theorem for (KL)-integral and [14, Theorem 3], it can be extended to a general $\sigma(P)$-measurable $x^*\mu_\infty$-integrable function ϕ. Hence $\phi.f$ is $(HKL)\mu_\infty$-integrable in T. Also if $E \in \sigma(P)$ and f is $(HKL)\mu_\infty$-integrable in T, i.e., if $(HKL)\int_E fd|x^*\mu_\infty| = x_E^{(HK)} \in \mathcal{X}$, then

$$\begin{aligned} x^*(x_E^{(HK)}) &= HK\int_E fd(|x^*\mu_\infty|) \\ &= HK\int_T \chi_E fd(|x^*\mu_\infty|) \end{aligned}$$

for $x^* \in \mathcal{X}^*$. □

Theorem 4.6 *If $u \in \mathcal{H}(\mathcal{X}, Y)$, then $u\mu_\infty : P \to \mathcal{X}$ is a σ-additive and f is (HKL)$u\mu_\infty$-integrable in T whenever f is (HKL) μ_∞-integrable in T. In that case*

$$u(HKL)\int_E fd\mu_\infty = (HKL)\int_E fd(u\mu_\infty) \textit{for } E \in \sigma(P).$$

Proof Since $u\mu_\infty : P \to \mathcal{X}$ is σ-additive, so $u\mu_\infty : P \to Y$ is σ-additive for $y^* \in Y^*$, $y^*u \in \mathcal{X}^*$, and hence we get the proof. □

4.4 μ_∞-a.e. Version of Convergences

Theorem 4.7 (Dominated Convergence Theorem of μ_∞-a.e. convergence version) *Dominated Convergence Theorem: Let f_n, $n \in \mathbb{N}$ be μ_∞-measurable with values in K or $[-\infty, \infty]$ and let $g : T \to K$ be (HKL) μ_∞-integrable in T. Suppose $f_n(t)| \le |g(t)|$ μ_∞-a.e. in T for all n. If $f : T \to K$ and if $f_n(t) \to f(t)$ μ_∞-a.e. in T, then f, f_n, $n \in \mathbb{N}$ are (HKL) μ_∞-integrable in T and*

$$(HKL)\int_E fd\mu_\infty = \lim_{n\to\infty}(HKL)\int_E f_n d\mu_\infty,\ E \in \sigma(P),$$

where the limit is uniform with respect to $E \in \sigma(P)$. Moreover,

$$\lim_{n\to\infty} \sup_{|x^*|\le 1} \int_T |f_n - f| d|\mu_\infty|(|x^*\mu_\infty|) = 0.$$

Proof Let us consider that all the μ_∞-measurable functions are $\sigma(P)$ measurable. Let $\nu(E) = HKL \int_E g d|x^*\mu_\infty|$, $E \in \sigma(P)$, then by Theorem 4.3(2), ν is σ-additive on $\sigma(P)$. Let f_n, $n \in \mathbb{N}$ be $(HKL)-\mu_\infty$-integrable in T. As $f_n \to f$ μ_∞-a.e. in T, then clearly f is $\overline{\sigma(P)}$ measurable. This means f is $\sigma(P)$-measurable. Moreover, $|f| \le |g|$ μ_∞-a.e. in T. From [32, Theorem 2.1.5(vii)], f is $(HKL)\mu_\infty$-integrable in T, then there exists $M \in \sigma(P)$ with $||\mu_\infty||(E) = 0$ such that g is finite on $T \setminus M$ and $f_n(t) \to f(t)$ for $t \in T \setminus M$. Let $F = \bigcup_{n=1}^{\infty} N(f_n) \cap (T \setminus M) = \bigcup_{n=1}^{\infty} (f_n \chi_{T\setminus M})$. Then $F \in \sigma(P)$ and $f_n \chi_{T\setminus M} \to f \chi_{T\setminus M}$ pointwise in T. Using the [11, Corollary 5], ν has a control measure $\mu_{\infty_0} : \sigma(P) \to [0, \infty)$. Thus for $\epsilon > 0$ there exists a $\delta > 0$ such that $\mu_{\infty_0}(E) < \delta$ implies $||\nu||(E) < \epsilon$. Now by Egoroff-Lusin theorem there exists a set $N \in \sigma(P) \cap F$ with $\mu_{\infty_0}(N) = 0$ and a sequence $\{F_k\}_{k=1}^{\infty} \subset P$ with F_k converges in $F \setminus N$ such that $f_n \to f$ uniformly in each F_k. As $\mu_{\infty_0}(F \setminus N \setminus F_k) \to 0$ there exists k_0 such that $\mu_{\infty_0}(F \setminus N \setminus F_k) < \delta$ for $k \ge k_0$. Since $\mu_{\infty_0}(N) = 0$, $||\nu||(N) = 0$ and $||\mu_\infty||(M) = 0$ therefore $||\nu||(M) = 0$ and since $f_n \to f$ uniformly in F_{k_0}, then there exists n_0 such that $||f_n - f||_{F_{k_0}} ||\mu_\infty||(F_{k_0}) < \epsilon \; \forall\, n \ge n_0$. For $E \in \sigma(P)$ and for $|x^*| \le 1$ with [14, Proposition 4 §8] with the fact $\mu_{\infty_0}(F \setminus N \setminus F_{k_0}) < \delta$ and $||\nu||(N) = ||\nu||(M) = 0$ we get

$$\begin{aligned}
&|x^*(HKL) \int_E f_n d\mu_\infty - x^*(HKL) \int_E f d\mu_\infty| \\
&= |HK \int_E (f_n - f) d(|x^*\mu_\infty|) \\
&\le HK \int_E |f_n - f| d|\mu_\infty|(|x^*\mu_\infty|) \\
&\le HK \int_{E\cap F_{k_0}} |f_n - f| d|\mu_\infty|(|x^*\mu_\infty|) \\
&+ HK \int_{E\cap(T\setminus F_{k_0})} |f_n - f| d|\mu_\infty|(|x^*\mu_\infty|).
\end{aligned}$$

Finally from [14, Proposition 8 §8] and the Theorem 4.3 we have

$$|(HKL) \int_E f_n d\mu_\infty - (HKL) \int_E f d\mu_\infty|$$

$$\leq \sup_{|x^*|\leq 1} HK \int_E |f_n - f| d|\mu_\infty|(|x^*\mu_\infty|)$$

$$\leq \epsilon$$

for $n \geq n_o$ and for all $E \in \sigma(P)$. □

Theorem 4.8 (Bounded Convergence Theorem of μ_∞-a.e. convergence version) *Bounded Convergence Theorem: Let P be a σ-ring. Let $f_n, n \in \mathbb{N}$, be μ_∞-measurable with values in K or in $[-\infty, \infty]$ and let K be a positive constant such that $|f_n(t)| \leq K$ μ_∞-a.e. in T for all n. If $f : T \to K$ and $f_n(t) \to f(t)$ μ_∞-a.e. in T, then f, $f_n, n \in \mathbb{N}$ are (HKL) μ_∞-integrable in T and*

$$(HKL) \int_E f dm = \lim_n (HKL) \int_E f_n dm \text{ for } E \in S,$$

the limit being uniform with respect to $E \in S$.

Proof Since P is a σ-ring and as the constant functions are (HKL) μ_∞-integrable in T so the proof follows as the Theorem 4.7. □

4.5 Set-Valued Kluvánek-Lewis-Henstock-Kurzweil Integrals

Let $HK(T, \mathbb{R}, |\mathcal{M}|)$ be the space of all functions $f : T \to \mathbb{R}$ which are Σ-measurable and $|\mathcal{M}|$-integrable.

Definition 4.6 Let $\mathcal{M} : \Sigma \to P_{wkc}(\mathcal{X})$ be a set-valued measure and $f : T \to \mathbb{R}$ an element of $HK(T, \mathbb{R}, |M|)$. Then f is said to be *Kluvánek-Lewis-Henstock-Kurzweil integrable* with respect to $\mathcal{M}$ (for short) $(KHL)\mathcal{M}$-integrable if

1. f is $\sigma(x^*, \mathcal{M}(.))$-integrable for each $x^* \in \mathcal{X}^*$.
2. For each $A \in \Sigma$, there exists a $W_A^{HK} \in P_{wkc}(\mathcal{X})$ such that

$$\sigma(x^*, W_A^{HK}) = HK \int_A f(w) d\sigma(x^*, \mathcal{M}(w))$$

for each $x^* \in \mathcal{X}^*$.

In this case, we write $W_A^{HK} = (HKL) \int_A f(w) d\mathcal{M}(w)$ for each $A \in \Sigma$ and call it set-valued *Kluvánek-Lewis-Henstock-Kurzweil integral* of f with respect to $\mathcal{M}$ on A.

Remark 4.2 The set-valued *Kluvánek-Lewis-Henstock-Kurzweil integral* is generalization of the *Kluvánek-Lewis-Henstock-Kurzweil integral*, when the set-valued measure $\mathcal{M}$ is degenerated into a vector-valued measure μ_∞, the set W_A^{HK} is reduced to a vector in $\mathcal{X}$, and the equality

$$\sigma(x^*, W_A^{HK}) = HK \int_A f(w) d\sigma(x^*, \mathcal{M}(w))$$

turns into

$$x^*(W_A^{HK}) = HK \int_A f(w) dx^* \mu_\infty(w)$$

for each $x^* \in \mathcal{X}^*$.

Example 4.1 Let $\mathcal{M} : \Sigma \to P_{wkc}(\mathcal{X})$ be a set-valued measure. If $f = \sum_{i=1}^{n} \mathcal{X}_i \xi_{A_i}$ is a nonnegative simple function, then f is $(HKL)\mathcal{M}$-integrable and

$$(HKL) \int_A f(w) d\mathcal{M}(w) = \sum_{i=1}^{n} \lambda_i (\mathcal{M}(A_i \cap A)) \tag{4.8}$$

for each $A \in \Sigma$.

Some properties of the set-valued *Kluvánek-Lewis-Henstock-Kurzweil integral* are given below:

Theorem 4.9 *Let $\mathcal{M} : \Sigma \to P_{wkc}(\mathcal{X})$ be a set-valued measure, $f, g : T \to \mathbb{R}(HKL)M$-integrable and $\lambda \geq 0$. Then*

1. *$f + g$ is $(HKL)\mathcal{M}$-integrable and*

$$(HKL) \int_A (f(w) + g(w)) d\mathcal{M}(w) = (HKL) \int_A f(w) d\mathcal{M}(w) + (HKL) \int_A g(w) d\mathcal{M}(w)$$

for each $A \in \Sigma$.

2. *λf is $(HKL)\mathcal{M}$-integrable and $(HKL) \int_A \lambda f(w) d\mathcal{M}(w) = \lambda (HKL) \int_A f(w) d\mathcal{M}(w)$ for each $A \in \Sigma$.*

Proof For (1) Since f and g are $(HKL)\mathcal{M}$-integrable, f and g are $\sigma(x^*, \mathcal{M}(.))$-integrable for each $x^* \in \mathcal{X}^*$ and there exist $(HKL) \int_A f(w) d\mathcal{M}(w)$, $(HKL) \int_A g(w) d\mathcal{M}(w)$ in $P_{wkc}(\mathcal{X})$ such that

$$\sigma(x^*, (HKL) \int_A f(w) d\mathcal{M}(w)) = \int_A f(w) d\sigma(x^*, \mathcal{M}(w))$$

$$\sigma(x^*, (HKL) \int_A g(w) d\mathcal{M}(w)) = \int_A g(w) d\sigma(x^*, \mathcal{M}(w))$$

for each $A \in \Sigma$ and $x^* \in \mathcal{X}^*$. Obviously, $f+g$ is $\sigma(x^*, \mathcal{M}(.))$-integrable and by properties of support functions, we have

$$\begin{aligned}
&\sigma(x^*, (HKL)\int_A f(w)d\mathcal{M}(w) + (HKL)\int_A g(w)d\mathcal{M}(w)) \\
&= \sigma(x^*, (HKL)\int_A f(w)d\mathcal{M}(w)) + \sigma(x^*, (HKL)\int_A g(w)d\mathcal{M}(w)) \\
&= HK\int_A f(w)d\sigma(x^*, \mathcal{M}(w)) + HK\int_A g(w)d\sigma(x^*, \mathcal{M}(w)) \\
&= HK\int_A \{f(w)+g(w)\}d\sigma(x^*, \mathcal{M}(w))
\end{aligned}$$

for each $A \in \Sigma$ and $x^* \in \mathcal{X}^*$.

Since $(HKL)\int_A f(w)d\mathcal{M}(w)$, $(HKL)\int_A g(w)d\mathcal{M}(w) \in P_{wkc}(\mathcal{X})$. It follows that $f+g$ is $(HKL)\mathcal{M}$-integrable and

$$\begin{aligned}
(HKL)\textstyle\int_A (f(w)+g(w))d\mathcal{M}(w) &= (HKL)\textstyle\int_A f(w)d\mathcal{M}(w) \\
&\quad + (HKL)\textstyle\int_A g(w)d\mathcal{M}(w)
\end{aligned}$$

for each $A \in \Sigma$.
For (2) : We have

$$\begin{aligned}
\sigma(x^*, \lambda(HKL)\int_A f(w)d\mathcal{M}(w)) &= \lambda\sigma(x^*, (HKL)\int_A f(w)d\mathcal{M}) \\
&= \lambda(HK)\int_A f(w)d\sigma(x^*, \mathcal{M}(w)) \\
&= HK\int_A \lambda f(w)d\sigma(x^*, \mathcal{M}(w)).
\end{aligned}$$

For each $x^* \in \mathcal{X}^*$, $A \in \Sigma$ and $\lambda \geq 0$. Since $(HKL)\int_A f(w)d\mathcal{M}(w) \in P_{wkc}(\mathcal{X})$, we have $\lambda(HKL)\int_A f(w)d\mathcal{M}(w) \in P_{wkc}(\mathcal{X})$. Thus λf is $(HKL)\mathcal{M}$-integrable and λf is $(HKL)\mathcal{M}$-integrable and

$(HKL)\int_A \lambda f(w)d\mathcal{M}(w) = \lambda(HKL)\int_A f(w)d\mathcal{M}(w)$ for each $A \in \Sigma$. □

Theorem 4.10 *Let (T, Σ, μ_∞) be a nonnegative finite measure space. $\mathcal{M} : \Sigma \to P_{wkc}(\mathcal{X})$ a μ_∞-continuous set-valued measure and $f : T \to \mathbb{R}$ $(HKL)\mathcal{M}$-integrable, then $\mathfrak{M} : \Sigma \to P_{wkc}(\mathcal{X})$ defined by $\mathfrak{M}(A) = (HKL)\int_A f(w)d\mathcal{M}(w)$, $\forall A \in \Sigma$ is μ_∞-continuous set-valued measure .*

Proof Clearly

$$\begin{aligned}\sigma(x^*, \mathfrak{M}(\emptyset)) &= \sigma(x^*, (HKL)\int_{\emptyset} f(w)d\mathcal{M}(w)) \\ &= HK\int_{\emptyset} f(w)d\sigma(x^*, \mathcal{M}(w)) \\ &= 0\end{aligned}$$

for each $x^* \in \mathcal{X}^*$. Also,

$$\begin{aligned}\sigma(x^*, \mathfrak{M}(A)) &= \sigma(x^*, (HKL)\int_{A} f(w)d\mathcal{M}(w)) \\ &= HK\int_{A} f(w)d\sigma(x^*, \mathcal{M}(w))\end{aligned}$$

If $\sigma(x^*, \mathcal{M}(.))$ is a μ_∞-continuous scalar-valued measure for each $x^* \in \mathcal{X}^*$, then for each $x^* \in \mathcal{X}^*$ and $A \in \Sigma$, we have

$$\begin{aligned}\sigma(x^*, \mathfrak{M}(A)) &= \sigma(x^*, (HKL)\int_{A} f(w)d\mathcal{M}(w)) \\ &= HK\int_{A} f(w)d\sigma(x^*, \mathcal{M}(w)).\end{aligned}$$

Since f is $\sigma(x^*, \mathcal{M}(w))$-integrable, then
$A \to (HK)\int_A f(w)d\sigma(x^*, \mathcal{M}(w))$ is countable additive. Let us assume $\{A_n\}_{n\in\mathbb{N}}$ be a sequence of pairwise disjoint elements of Σ, then

$$\begin{aligned}\sigma(x^*, \mathfrak{M}(\cup_{n=1}^{\infty} A_n) &= \sigma(x^*, (HKL)\int_{\cup_{n=1}^{\infty} A_n} f(w)d\mathcal{M}(w)) \\ &= HK\int_{\cup_{n=1}^{\infty} A_n} f(w)d\sigma(x^*, \mathcal{M}(w)) \\ &= \sum_{n=1}^{\infty} HK\int_{A_n} f(w)d\sigma(x^*, \mathcal{M}(w)) \\ &= \sum_{n=1}^{\infty} \sigma(x^*, (HKL)\int_{A_n} f(w)d\mathcal{M}(w)) \\ &= \sum_{n=1}^{\infty} \sigma(x^*, \mathfrak{M}(A_n)).\end{aligned}$$

Therefore $\sigma(x^*, \mathfrak{M}(.))$ is a scalar-valued measure for each $x^* \in \mathcal{X}^*$. Hence $\mathfrak{M} : \Sigma \to P_{wkc}(\mathcal{X})$ is a continuous set-valued measure. □

Theorem 4.11 *Suppose $\mathcal{X}$ has the Radon Nikodým Property (RNP), $\mathcal{X}^*$ is separable, $\mathcal{M} : \Sigma \to P_{wkc}(\mathcal{X})$ is a μ_∞-continuous set-valued measure of bounded variation and $f : T \to \mathbb{R}^+$ is $(HKL)\mathcal{M}$-integrable, then there exists an integrably bounded set-valued function $X : T \to P_{wkc}(\mathcal{X})$ such that*

$$(HKL) \int_T f(w) d\mathcal{M}(w) = HK \int_T f(w)\mathcal{X}(w) d\mu_\infty(w).$$

Proof Since $\mathcal{X}$ has the RNP, $\mathcal{X}^*$ is separable, $\mathcal{M} : \Sigma \to P_{wkc}(\mathcal{X})$ is a μ_∞-continuous set-valued measure of bounded variation, then there exists an integrably bounded set-valued function $\mathcal{X} : T \to P_{wkc}(\mathcal{X})$ which is a Radon Nikodým derivative of $\mathcal{M}$ with respect to μ_∞, then $\mathcal{M}(A) = HK \int_A \mathcal{X}(w) d\mu_\infty(w)$ for each $A \in \Sigma$. It follows that

$$\begin{aligned} \sigma(x^*, \mathcal{M}(A)) &= \sigma(x^*, HK \textstyle\int_A \mathcal{X}(w) d\mu_\infty(w)) \\ &= HK \textstyle\int_A \sigma(x^*, \mathcal{X}(w)) d\mu_\infty(w) \end{aligned} \tag{4.9}$$

for each $x^* \in \mathcal{X}^*$ and each $A \in \Sigma$.

Now from the fact that $\mathcal{M} : \Sigma \to P_{wkc}(\mathcal{X})$ is μ_∞-continuous set-valued measure iff $\sigma(x^*, \mathcal{M}(.))$ is μ_∞-continuous scalar-valued measure for each $x^* \in \mathcal{X}^*$. This fact with (4.9) gives $\sigma(x^*, \mathcal{X}(w))$ is a Radon-Nikodým derivative of $\sigma(x^*, \mathcal{M}(.))$ with respect to μ_∞ for each $x^* \in \mathcal{X}^*$. Therefore

$$\begin{aligned} &\sigma(x^*, (HKL) \textstyle\int_T f(w) d\mathcal{M}(w)) \\ &= HK \textstyle\int_T f(w) d\sigma(x^*, \mathcal{M}(w)) \\ &= HK \textstyle\int_T f(w)\sigma(x^*, \mathcal{X}(w)) d\mu_\infty(w) \end{aligned} \tag{4.10}$$

for each $x^* \in \mathcal{X}^*$. Since f is nonnegative from [41, Theorem 8], we have

$$\begin{aligned} &HK \int_T f(w)\sigma(x^*, \mathcal{X}(w)) d\mu_\infty(w) \\ &= HK \int_T \sigma(x^*, f(w)\mathcal{X}(w)) d\mu_\infty(w) \\ &= \sigma(x^*, HK \int_T f(w)\mathcal{X}(w) d\mu_\infty(w) \end{aligned}$$

for each $x^* \in \mathcal{X}^*$. From (4.10)

$$(HKL) \int_T f(w) d\mathcal{M}(w) = HK \int_T f(w)\mathcal{X}(w) d\mu_\infty(w)$$

□

Theorem 4.12 *Let $\mathcal{M} : \Sigma \to P_{wkc}(\mathcal{X})$ be a set-valued measure and let $f : T \to \mathbb{R}$ be $(HKL)\mathcal{M}$-integrable, then*

$$|(HKL)\int_A f(w)d\mathcal{M}(w)| \leq HK\int_A |f(w)|d(w)$$

for each $A \in \Sigma$.

Proof As f is $(HKL)\mathcal{M}$-integrable, there exists $(HKL)\int_A f(w)d\mathcal{M}(w) \in P_{wkc}(\mathcal{X})$ such that

$$\sigma(x^*, (HKL)\int_A f(w)d\mathcal{M}(w)) = HK\int_A f(w)d\sigma(x^*, \mathcal{M}(w))$$

for each $A \in \Sigma$ and $x^* \in \mathcal{X}^*$. From the fact of support function

$$\begin{aligned}
&|(HKL)\int_A f(w)d\mathcal{M}(w)| \\
&= H((HKL)\int_A f(w)d\mathcal{M}(w), \{0\}) \\
&= \sup_{||x^*||\leq 1} |\sigma(x^*, (HKL)\int_A f(w)d\mathcal{M}(w)| \\
&= \sup_{||x^*||\leq 1} |HK\int_A f(w)d\sigma(x^*, \mathcal{M}(w)| \\
&\leq \sup_{||x^*||\leq 1} (HK)\int_A |f(w)|d|\sigma(x^*, \mathcal{M}(.)|(w).
\end{aligned}$$

Now,

$$\begin{aligned}
|\sigma(x^*, \mathcal{M}(0))|(A) &= \sup\sum_{i=1}^{n} |\sigma(x^*, \mathcal{M}(A_i)| \\
&\leq \sup\sum_{i=1}^{n} ||x^*|||\mathcal{M}(A_i)| \\
&\leq \sup\sum_{i=1}^{n} |\mathcal{M}(A_i)| \\
&= |\mathcal{M}|(A)
\end{aligned}$$

for each $A \in \Sigma$ and $x^* \in \mathcal{X}^*$ with $||x^*|| \leq 1$, where the supremum is taken over all finite measurable partition $\{A_1, A_2, \ldots, A_n\}$ of A, as f is nonnegative so

$$HK \int_A |f(w)|d|\sigma(x^*, \mathcal{M}(.))|(w) \leq HK \int_A |f(w)d|\mathcal{M}|(w)$$

for each $A \in \Sigma$ and $x^* \in \mathcal{X}^*$ with $||x^*|| \leq 1$. Therefore

$$\begin{aligned}
&|(HKL) \int_A f(w)d\mathcal{M}(w)| \\
&\leq \sup_{||x^*|| \leq 1} HK \int_A \int_A |f(w)|d|\sigma(x^*, \mathcal{M}(w))| \\
&\leq HK \int_A |f(w)|d|\mathcal{M}|(w) \\
&\leq HK \int_A |f(w)|d(w)
\end{aligned}$$

for each $A \in \Sigma$. □

Corollary 4.1 *Let $\mathcal{M} : \Sigma \to P_{wkc}(\mathcal{X})$ be a set-valued measurable and let $f : T \to \mathbb{R}$ be $(HKL)\mathcal{M}$-integrable, then*

1. $\lim_{A \to \emptyset} H((HKL) \int_A f(w)d\mathcal{M}(w), \{0\}) = 0.$
2. *If $f = \emptyset$, then $(HKL) \int_A f(w)\mathcal{M}(w) = \{0\}$.*
3. *If $f = 0$, then $(HKL) \int_A f(w)d\mathcal{M}(w) = \{0\}$*

for each $A \in \Sigma$.

Theorem 4.13 *Let $\mathcal{M} : \Sigma \to P_{wkc}(\mathcal{X})$ be a set-valued measure and let $f, f_n : T \to \mathbb{R}, n \in \mathbb{N}$ be $(HKL)\mathcal{M}$-integrable such that $\{f_n\}_{n\in\mathbb{N}}$ is uniformly integrable with respect to $|\mathcal{M}|$ and $\lim_{n\to\infty} f_n = f$, $|\mathcal{M}|$-a.e., then*

$$\lim_{n\to\infty} H((HKL) \int_T f_n(w)d\mathcal{M}(w), (HKL) \int_T f(w)d\mathcal{M}(w)) = 0.$$

Proof Given $f, f_n : T \to \mathbb{R}, n \in \mathbb{N}$ are $(HKL)\mathcal{M}$-integrable, then there exist $(HKL) \int_T f_n(w)d\mathcal{M}(w)$, $(HKL) \int_T f(w)d\mathcal{M}(w) \in P_{wkc}(\mathcal{X})$, $n \in \mathbb{N}$ such that

$$\sigma(x^*, (HKL) \int_T f_n(w)d\mathcal{M}(w)) = HK \int_T f_n(w)d\sigma(x^*, \mathcal{M}(w))$$

and

$$\sigma(x^*, (HKL) \int_T f(w)d\mathcal{M}(w)) = HK \int_T f(w)d\sigma(x^*, \mathcal{M}(w))$$

for each $x^* \in \mathcal{X}^*$ and $n \in \mathbb{N}$. The properties of support function give:

$$\begin{aligned}
&H((HKL)\int_T f_n(w)d\mathcal{M}(w), (HKL)\int_T f(w)d\mathcal{M}(w)) \\
&= \sup_{||x^*||\le 1} |\sigma(x^*, (HKL)\int_T f_n(w)d\mathcal{M}(w)) \\
&\quad - \sigma(x^*, (HKL)\int_T f(w)d\mathcal{M}(w))| \\
&= \sup_{||x^*||\le 1} |HK\int_T f_n(w)d\sigma(x^*, \mathcal{M}(w)) \\
&\quad - HK\int_T f(w)d\sigma(x^*, \mathcal{M}(w)| \\
&\le \sup_{||x^*||\le 1} HK\int_T |f_n(w) - f(w)|d|\sigma(x^*, \mathcal{M}(.))|(w) \\
&\le HK\int_T |f_n(w) - f(w)|d|\mathcal{M}|(w).
\end{aligned}$$

Now using Vitali's convergence theorem

$$\begin{aligned}
&H((HKL)\int_T f_n(w), d\mathcal{M}(w), (HKL)\int_T f(w)d\mathcal{M}(w)) \\
&\le HK\int_T |f_n(w) - f(w)|d|\mathcal{M}|(w) \ \to 0 \, as \; n \to \infty.
\end{aligned}$$

□

We can construct (KL)-type dominated convergence theorem for the set-valued Kluvánek-Lewis-Henstock-Kurzweil integral as follows:

Theorem 4.14 *Let $\mathcal{M} : \Sigma \to P_{wkc}(\mathcal{X})$ be a set-valued measure and let $f, f_n : T \to \mathbb{R}, n \in \mathbb{N}$ be $(HKL)\mathcal{M}$-integrable such that $\lim_{n\to\infty} f_n = f|\mathcal{M}|$-a.e. If there exists a nonnegative $|\mathcal{M}|$-integrable function $g : T \to \mathbb{R}$ such that $|f_n(w)| \le g(w)$ for all $n \in \mathbb{N}$ and $w \in T$, then*

$$\lim_{n\to\infty} H((HK)\int_T f_n(w)d\mathcal{M}(w), (HK)\int_T f(w)d\mathcal{M}(w)) = 0.$$

Proof Proof is similar as [41, Theorem 12]. □

References

1. Boccuto, A., Candeloro, D., Sambucini, A.R.: Henstock multivalued integrability in Banach lattices with respect to pointwise non atomic measures. Rendiconti Lincei-Matematica e Applicazioni **26**(4), 363–383 (2015)
2. Boccuto, A., Hazarika, B., Kalita, H.: Kuelbs-Steadman spaces for Banach space-valued measures. Mathematics **8**(6), 1–12 (2020)
3. Boccuto, A., Riečan, B., Vrábelová, M.: Kurzweil-Henstock Integral in Riesz Spaces. Bentham Science Publisher, Sharjah (2009)
4. Brézis, H.: Functional Analysis, Sobolev Spaces and Partial Differential Equations. Springer, New York-Dordrecht-Heidelberg-London (2011)
5. Candeloro, D., Di Piazza, L., Musiał, K., Sambucini, A.R.: Gauge integrals and selections of weakly compact valued multifunctions. J. Math. Anal. Appl. **441**(1), 293–308 (2016)
6. Candeloro, D., Di Piazza, L., Musiał, K., Sambucini, A.R.: Some new results on integration for multifunction. Ricerche di Matematica **67**(2), 361–372 (2018)
7. Curbera, G.P., Ricker, W.J.: Vector Measures, Integration and Applications, Trends in Mathematics, Positivity, pp. 127–160 (2007)
8. Candeloro, D., Sambucini, A.R.: Comparison between some norm and order gauge integrals in Banach lattices. Panamerican Math. J. **25**(3), 1–16 (2015)
9. Candeloro, D., Sambucini, A.R.: Order-type Henstock and McShane integrals in Banach lattice setting. In: Proceedings of the SISY 2014 - IEEE 12th International Symposium on Intelligent Systems and Informatics, pp. 55–59 (2014)
10. Cao, S.: The Henstock integral for Banach-valued functions. Southeast Asian Bull. Math. **16**(1), 35–40 (1992)
11. Diestel, J., Uhl, J.J.: Vector measures. American Mathematical Society, Providence (1977)
12. Di Piazza, L.: Kurzweil-Henstock type integration on Banach spaces. Real Anal. Exchange **29**(2), 543–555 (2003/2004)
13. Di Piazza, L., Musiał, K.: Set valued Kurzweil-Henstock-Pettis integral. Set-Valued Anal. **13**, 167–169 (2005)
14. Dinculeanu, N.: Vector Measures, vol. 95 of International Series of Monographs in Pure and Applied Mathematics. Pergamon Press, New York (1967)
15. Dobrakov, I., Panchapagesan, T.V.: A generalized Pettis measurability criterion and integration of vector functions. Studia Mathematica **164**, 205–229 (2004)
16. Džamonja, M., Kunen, K.: Properties of the class of measure separable compact spaces. Fundamenta Mathematicae **147**, 261–277 (1995)
17. Federson, M.: Some Peculiarities of the HK-integrals of Banach space-valued functions. Real Anal. Exchange **29**(1), 439–460 (2003/2004)
18. Fremlin, D.H., Mendoza, J.: On the integration of vector-valued functions. Illinois J. Math. **38**(1), 127–147 (1994)
19. Gordon, R.: The Integrals of Lebesgue, Denjoy, Perron, and Henstock. American Mathematical Society, Providence (1994)
20. Kalita, H., Hazarika, B.: Countable additivity of Henstock-Dunford integrable functions and Orlicz space. Anal. Math. Phys. **11**(2), 1–13 (2021)
21. Kalita, H., Hazarika, B.: Kluvánek-Lewis-Henstock integral in a Banach space. Bollettino dell'Unione Matematica Italiana **17**, 677–695 (2024)
22. Kandilakis, D.A.: On the extension of multimeasures and integration with respect to a multimeasure. Proc. Am. Math. Soc. **116**(1), 85–92 (1992)
23. Kluvánek, I., Knowles, G.: Vector Measures and Control Systems. Mathematics Studies. North Holland Publishing Company, Amsterdam (1975)
24. Kluvánek, I.: Fourier transforms of vector-valued functions and measures. Studia Mathematica **37**, 1–12 (1970)
25. Lee, P.Y., Výborný, R.: The Integral: An Easy Approach After Kurzweil and Henstock. Cambridge University Press, Cambridge (2000)

26. Lewis, D.R.: On integrability and summability in vector spaces. Illinois J. Math. **16**, 294–307 (1972)
27. Lewis, D.R.: Integration with respect to vector measures. Pac. J. Math. **33**(1), 157–165 (1970)
28. Lindenstrauss, J., Tzafriri, L.: Classical Banach Spaces II, Function Spaces. A Series of Modern Surveys in Mathematics. Springer, Berlin-Heidelberg-New York (1979)
29. McLeod, R.M.: The Generalized Riemann Integral. Carus Mathematical Monograph, The Mathematical Association of America 20, New York (1980)
30. Okada, S.: The dual space of $\mathcal{L}^1(\mu)$ for a vector measure μ. J. Math. Anal. Appl. **177**, 583–599 (1993)
31. Papageorgiou, N.S.: On the theory of Banach space valued multifunctions. 2. Set valued martingales and set valued measures. J. Multivariate Anal. **17**(2), 207–227 (1985)
32. Panchapagesan, T.V.: The Bartle-Dunford-Schwartz Integral–Integration with Respect to a Sigma-Additive Vector Measure. Birkhäuser, Basel Boston-Berlin (2008)
33. Pfeffer, W.F.: The multidimensional fundamental theorem of Calculus. J. Aust. Math. Soc. A **43** , 143–170 (1987)
34. Precupanu, A.M., Croitoru, A.: A Gould type integral with respect to a multimeasure I. Analele Universitatii Alexandru Ioan Cuza Iaşi, (New Series) Mathematics XLVIII., **1**, 165–200 (2002)
35. Precupanu, A.M., Croitoru, A.: A Gould type integral with respect to a multimeasure II. Analele Universitatii Alexandru Ioan Cuza Iaşi XLIX 183–207 (2003)
36. Ricker, W.: Separability of the L^1-space of a vector measure. Glasgow Math. J. **34**(1), 1–9 (1992)
37. Riečan, B.: On the Kurzweil integral in compact topological spaces. Radovi Matematicki **2**, 151–163 (1986)
38. Roy, S.K., Chakraborty, N.D.: On Orlicz spaces for a family of measures. J. Math. Anal. Appl. **145**(2), 485–503 (1990)
39. Swartz, C.: Introduction to Gauge Integral. World Scientific, Singapore (2001)
40. Wu, W.Z., Zhang, W.X., Wang, R.-M.: Set valued Bartle integrals. J. Math. Anal. Appl. **255**(1), 1–20 (2001)
41. Zhou, C.L. and Shi, F.G.: New set-valued integral in Banach space. J. Funct. Spaces **2015**(1), 260238 (2015)
42. Zhang, W., Li, S., Wang, Z., Gao, Y.: Set Valued Stochastic Process. Academic Press, Beijing (2007)

Chapter 5
Henstock-Dunford and Henstock-Pettis Integrable Functions

In this chapter we investigate the space of the Henstock-Dunford integral function. For each function which is Henstock-Dunford integral can be constructed by an operator. For any Henstock-Dunford integrable function f, an operator can be constructed such that it is both a continuous linear operator and a weakly compact operator. Furthermore, the adjoint operator is continuous and weakly compact linear operator for the case of Henstock-Dunford integral. In the end, we introduce Pettis integrability-type property for HKP-integrals. We discuss several necessary conditions that $\mathcal{X}$ has HKP-integrability property for weak Baire measure. The necessary and sufficient conditions is that the indefinite integral of any Henstock-Kurzweil-Pettis (respectively, Denjoy-Pettis) integrable function with values in a fixed Banach space have separable ranges are discussed.

5.1 Introduction

In 1912, A. Denjoy developed an integration process that satisfies:
Let $F : [a, b] \to \mathbb{R}$ be continuous. If F is differentiable nearly everywhere on $[a, b]$, then F' is integrable on $[a, b]$ and $\int_a^x F' = F(x) - F(a)$ for each $x \in [a, b]$.

Denjoy called the process of computing the value of his integral "totalization" and showed that every derivative met the criteria for this process and that original function was recovered. That is, Denjoy integral is a simple generalization of the characterization of the Lebesgue integral. It is well known that the Henstock-Kurzweil integral is more powerful and simpler than the Lebesgue and Feynman integrals, see [11, 13]. Candeloro et al. [6] recently studied about the integration theory and the multifunctions. It has played an important role in applications to several branches of science, like control theory, differential inclusions, game theory, aggregation functions, economics, problems of finding equilibria, and optimization. Since neither the Riemann integral nor the Lebesgue integral is

B. Hazarika, H. Kalita, *Theory of Henstock–Orlicz Spaces*, Mathematical Marvels: Texts and Monographs in the Spirit of CR Rao,
https://doi.org/10.1007/978-981-96-9548-5_5

completely satisfactory concerning the problem of the existence of primitives, different types of integrals extending the previous ones have been introduced and investigated, like Henstock-Kurzweil, McShane, and Pettis integrals. These topics have many connections with measures taking values in abstract spaces, and in particular, the extension of the concept of integrability to set-valued functions can be used in order to obtain a larger number of selections for multifunctions, through their estimates and properties, in several applications, see [6, 7, 9, 10]. Henstock integral has been applied in different areas of functional analysis by Prez-Becerra et al. [19]. Gordon in [12] gave two Denjoy-type extensions of the Dunford, Pettis integrals, and the Denjoy-Dunford and Denjoy-Pettis integrals and discussed their properties. The Henstock-Dunford integral is generalized in the Euclidean space, namely, the Henstock-Dunford integral in the Euclidean space. Guoju and Tianqing in [14] gave the Riemann-type extensions of Dunford, Pettis integrals, the Henstock-Dunford, and Henstock-Pettis integrals and discussed the relationships between the Henstock-Dunford integral and Dunford integral, Henstock-Pettis integral, and Pettis integral. They proved the Harnack Extension Theorems and the convergence theorems for Henstock-Dunford and Henstock-Pettis integrals. Ju Han Yoon, Jae Myung Park, Young Kuk Kim, and Byung Moo Kim studied about the AP-Henstock extension of Dunford, Pettis, and Bochner integrals of functions mapping $[a, b]$ into Banach space $\mathcal{X}$, see [25]. The study of integral theory is also combined with operator theory. Several properties of the positive Dunford-Pettis operator have been discussed in [2]. Recently [16], we have shown the countable additivity of Henstock-Dunford integrable functions in Henstock-Orlicz spaces (in short H-Orlicz spaces) with the help of de la Vallèe Poussin Theorem [21, Theorem 2, page 3] and Dunford-Pettis Theorem. Detailed about Orlicz space and H-orlicz spaces one can see [15, 16, 21]. Piazza and Musiał' characterized Henstock-Kurzweil-Pettis integrals by cores of the functions and through the properties of suitable operations defined by integrands. Bongiorno et al. studied about Henstock-Kurzweil-Pettis integrable function, and Henstock integrable functions are scalarly approximated in the Alexiewicz norm by a sequence of step functions. In their work the authors have done the full approximation of the Henstock-Kurzweil-Pettis integral if and only if the range of the integral is norm relatively compact.Naralenkov found the necessary and sufficient conditions so that the indefinite integral of any Henstock-Kurzweil-Pettis integrable function with values in a fixed Banach space has a relatively compact range or is continuous, except at most on a countable set.

Operators work in the space of Bochner integrable function. Next, operators are examined in the space of Dunford integrable function. Solikhin et al. [23] concluded that the collection of all functions integrated Dunford is a linear space. For each function which is Dunford integral can be constructed by an operator. It is continuous linear and weakly compact operator. Furthermore, the adjoint operator is continuous and weakly compact linear operator.

5.2 Preliminaries and Definitions

Definition 5.1 ([12, Definition 8]) Let $f : [a, b] \to \mathcal{X}$ and $t \in (a, b)$. A vector $z \in \mathcal{X}$ is the approximate derivative of f if there exists a measurable set $E \subset [a, b]$ that has t as a point of density such that $\lim_{s \to t} \dfrac{f(s) - f(t)}{s - t} = z$ with $s \in E$. We denote it by F'_{ap}.

Definition 5.2 ([3]) A subset of a (real or complex) vector space is called absolutely convex (AC) provided it is convex and closed under scalar multiplication by scalars of absolute value less than equal to 1.

Definition 5.3 ([5, Definition 5.25]) Let $f : [a, b] \to \mathbb{R}$. We say that f is absolutely continuous if for each $\varepsilon > 0$ there exists $\varepsilon > 0$ such that if $\{[a_n, b_n]\}$ is any finite or countable collection of nonoverlapping closed intervals in $[a, b]$, with $\sum_{k=1}^{\infty}(b_k - a_k) < \infty$, then $\sum_{k=1}^{\infty} |f(b_k) - f(a_k)| < \epsilon$.

Definition 5.4 ([12, Definition 7.1]) A function $f : [a, b] \to \mathbb{R}$ is Denjoy integrable on $[a, b]$ if there exists an ACG_* function $F : [a, b] \to \mathbb{R}$ such that $F' = f$ a.e. on $[a, b]$.

Definition 5.5 ([14, 22]) A function $f : [a, b] \to \mathcal{X}$ is said to be Dunford integrable on $[a, b]$ if for each $x^* \in \mathcal{X}^*$, the real function $x^* f$ is Lebesgue integrable and for each set of measurable $A \subset [a, b]$ there exists a unique vector $x^{**}_{(f,A)} \in \mathcal{X}^{**}$ such that $x^{**}_{(f,A)}(x^*) = \int_A x^* f$, for every $x^* \in \mathcal{X}^*$.

Remark 5.1 (Guoju and Tianqing [14] Defined Henstock-Dunford Integral) Lemma 5.1 was needed for existence of the definition of Henstock-Dunford integral.

The set of all Dunford integrable functions is denoted by $D[a, b]$. For $f \in D[a, b]$ means that f is Dunford integrable on $[a, b]$.

Theorem 5.1 ([23, Theorem 3.4]) *The collection of all functions f that Dunford integrable on $[a, b]$, then $D[a, b]$ is linear space.*

Lemma 5.1 *Assume that $f : I = [a, b] \to \mathcal{X}$ is weakly measurable and that for each $x^* \in \mathcal{X}^*$, the function $x^*(f) : I = [a, b] \to \mathbb{R}$ is Henstock integrable. Then for each measurable $E \subset I = [a, b]$ there exists a unique $x^{**}_E \in \mathcal{X}^{**}$ such that $\int_E x^*(f) = x^{**}_E$ for every $x^* \in \mathcal{X}^*$.*

Proof For a given measurable $E \subset I$ we have $\int_E x^*(f) = \int_I x^*(f.\chi_E)$ and we can define $T_E(x^*) = x^*(f.\chi_E)$. T_E is a linear map of $\mathcal{X}^*$ into the space $HK(I)$ of Henstock-Kurzweil-integrable functions on I and $F_E : x^* \to \int_I x^*(f.\chi_E)$ is a linear functional on $\mathcal{X}^*$. Assume that $x^*_n \to x^*$ in $\mathcal{X}^*$ and $T_E(x^*_n) \to g$ in $HK(I)$ for $n \to \infty$. That is, $\lim_{n\to\infty} | \int_I x^*_n(f.\chi_E) - \int g| < \epsilon$. Then $x^*_n(f.\chi_E)$ converges for $n \to \infty$ in measure to g and by the Riesz Theorem there is a subsequence $x^*_{n_k}$, $k \in \mathbb{N}$ of x^*_n such that $x^*_{n_k}(f(t).\chi_E(t)) \to x^*(f(t).\chi_E(t))$, $n \to \infty$ for all $t \in I$. Since

$x_n^*(f(t).\chi_E(t)) \to x^*(f(t).\chi_E(t))$ for all $t \in I$. Consequently, the graph of the linear map $T_E : \mathcal{X}^* \to HK(I)$ is closed and by the Closed Graph Theorem the operator T_E is bounded. That is,

$$\begin{aligned}|F_E(x^*)| &= |\int_E x^*(f.\chi_E)| \\ &\leq \int_I |x^*(f.\chi_E)| \\ &= \int_I |T_E(x^*)| \\ &= ||T_E(x^*)||_{HK(I)} \\ &\leq ||T_E||.||||x^*||.\end{aligned}$$

Therefore $\int_E x^*(f)$ is a continuous linear functional on $\mathcal{X}^*$ defining an element $x_E^{**} \in \mathcal{X}^{**}$ for which $\int_E x^*(f) = x_E^{**}(x^*)$. □

Recalling a function $f : [a, b] \to \mathcal{X}$ is said to be Henstock-Dunford integrable on $[a, b]$ if for each $x^* \in \mathcal{X}^*$, the real function $x^* f$ is Henstock-Kurzweil integrable and for each set $A \subset [a, b]$ there exists a unique vector $x_{(f,A)}^{**} \in \mathcal{X}^{**}$ such that $x_{(f,A)}^{**}(x^*) = \int_A x^* f$, for every $x^* \in \mathcal{X}^*$, see [22].

We denote the set of all Henstock-Dunford integrable functions by $D_{HD}[a, b]$. That is for $f \in D_{HD}[a, b]$ means f is Henstock-Dunford integrable on $[a, b]$.

Theorem 5.2 *A function f is Henstock-Dunford integrable on $[a, b]$ if and only if the function $x^* f$ is Henstock integrable on $[a, b]$.*

Theorem 5.3 *The dual of $HK[a, b]$ are the class of functions of bounded essential variation ($BV_H[a, b]$). That is, $HK^*[a, b] = BV_H[a, b]$.*

5.3 The Space of Henstock-Dunford Integral Functions

In this section we discuss about $D_{HD}[a, b]$, the space of Henstock-Dunford integral function. We describe a few of the basic properties of the space and give references to some of the more technical properties.

Proposition 5.1 *If $f \in D_{HD}[a, b]$, then for every single measurable set $A \subset [a, b]$, vector $x_{(f,A)}^{**} \in \mathcal{X}^{**}$ is single.*

Proof Given any measurable set $A \subset [a, b]$, suppose there is a vector $x_{1(f,A)}^{**}$ and $x_{2(f,A)}^{**} \in \mathcal{X}^{**}$, then $x_{1(f,A)}^{**}(x^*) = HK \int_A x^*(f)$ and $x_{2(f,A)}^{**}(x^*) = HK \int_A x^*(f)$.

Therefore,

$$x^{**}_{1(f,A)}(x^*) - x^{**}_{2(f,A)}(x^*) = HK\int_A x^*(f) - HK\int_A x^*(f)$$
$$= 0, \forall x^* \in \mathcal{X}^*.$$

So, $x^{**}_{1(f,A)}(x^*) = x^{**}_{2(f,A)}(x^*)$. This completes the proof. □

Theorem 5.4 *The collection of all functions f that Henstock-Dunford integrable on $[a, b]$, then $D_{HD}[a, b]$ is linear space.*

Proof Let $f, g \in D_{HD}[a, b]$ arbitrarily and $c \in \mathbb{R}$. If $f, g \in D_{HD}[a, b]$, then for each $x^* \in \mathcal{X}^*$ the function x^*f & x^*g are Henstock-Kurzweil-integrable on $[a, b]$, so that for any $x^* \in \mathcal{X}^*$ the function $x^*(f+g) = x^*f + x^*g$ is Henstock-Kurzweil integrable on $[a, b]$. So, $f + g \in D_{HD}[a, b]$. Further $x^*(cf) = cx^*f$ is Henstock-Kurzweil integrable on $[a, b]$. Thus the function cf is Henstock-Dunford integrable on $[a, b]$. □

Theorem 5.5 *The step functions are dense in $D_{HD}[a, b]$.*

Proof Let $f \in D_{HD}[a, b]$ be a step function and $\epsilon > 0$. Definition of Henstock-Dunford integral shows $x^*f \in HK[a, b]$. Picking a gauge γ on $I = [a, b]$ such that

$$\left|S(x^*f, D) - \int_{[a,b]} x^*f\right| < \epsilon$$

whenever $D << \gamma$. Let $D = \{(t_i, I_i) : 1 \le i \le m\} << \gamma$ and $\theta = \sum\limits_{i=1}^{m} x^*f(t_i)C_{I_i}$. By Uniform Henstock Lemma if $J \in I = [a, b]$,

$$\left|\int_I x^*f - \int_J \theta\right| = \left|\sum_{i=1}^{m}\left\{\int_{J\cap I_i} x^*f - \int_{J\cap I_i}\theta\right\}\right|$$
$$= \left|\sum_{i=1}^{m}\left\{\int_{J\cap I_i} x^*f - x^*f(t_i)l(I_i \cap J)\right\}\right|$$
$$\le \epsilon.$$

Hence, $||x^*f - \theta|| \le \varepsilon$. Finally, we establish a sufficient condition for convergence of a sequence x^*f_n with respect the Alexiewicz norm. □

Theorem 5.6 *$BV[a, b]$ is subset of $D_{HD}[a, b]$.*

Proof Proof is straightforward using the concept of $BV[a, b]$. □

Theorem 5.7 *$D[a, b]$ is subspace of $D_{HD}[a, b]$.*

Proof Proof is easy so we left to the readers. □

5.4 Weakly Compact Operator on $D_{HD}[a, b]$

Let $\mathcal{X}$ be a Banach space with its dual $\mathcal{X}^*$ and $\mathcal{X}^{**}$ is dual of $\mathcal{X}^*$. Let us denote $HK[a, b]$ the space of the Henstock integrable functions on $[a, b]$.

Definition 5.6 We define the operator:

$$D_{HD} : X^* \to HK[a, b] \text{ by } D_{HD}(x^*) = x^* f \text{ for every } x^* \in \mathcal{X}^*.$$

This operator is well defined.

Theorem 5.8 *The operator D_{HD} is a bounded linear.*

Proof Let $x^*, t^* \in \mathcal{X}^*$ and $c \in \mathbb{R}$ be any scalar. Then

$$\begin{aligned} D_{HD}(x^* + t^*) &= (x^* + t^*) f \\ &= x^* f + t^* f \\ &= D_{HD}(x^*) + D_{HD}(x^*). \end{aligned}$$

And $D_{HD}(cx^*) = (cx^*) f = cx^* f = cD_{HD}(x^*)$. So, it is a linear operator. Furthermore, according to the Closed Graph Theorem, the operator D_{HD} is bounded. □

Definition 5.7 For each $f \in D_{HD}[a, b]$, we define $D^*_{HD} : HK^*[a, b] \to \mathcal{X}^{**}$ by an operator:

$$D^*_{HD}(h)(x^*) = \int_a^b h D_{HD}(x^*) = \int_a^b h x^* f$$

for every $h \in HK^*[a, b]$.

An operator T^* is called adjoint operator against the operator $T : \mathcal{X}^* \to HK[a, b]$ on $HK[a, b]$.

Theorem 5.9 *The adjoint operator D^*_{HD} is a bounded operator and $||D^*_{HD}|| = ||D_{HD}||$.*

Proof Let $h_1, h_2 \in BV_H[a, b]$ and any scalars $c_1, c_2 \in \mathbb{R}$. Clearly $c_1h_1 + c_2h_2 \in BV_H[a, b]$. Then

$$\begin{aligned}
&D^*_{HD}(c_1h_1 + c_2h_2)(x^*)\\
&= \int_a^b (c_1h_1 + c_2h_2)D_{HD}(x^*)\\
&= c_1 \int_a^b h_1D_{HD}(x^*) + c_2 \int_a^b h_2D_{HD}(x^*)\\
&= c_1D^*_{HD}(h_1)(x^*) + c_2D^*_{HD}(h_2)(x^*).
\end{aligned}$$

□

Theorem 5.10 *If $D_{HD} : \mathcal{X}^* \to HK[a, b]$ and $T_{HD} : \mathcal{X}^* \to HK[a, b]$ are bounded operators and any scalars $c \in \mathbb{R}$, then*

1. $(D_{HD} + T_{HD})^* = D^*_{HD} + T^*_{HD}$.
2. $(cD_{HD})^* = cD^*_{HK}$.
3. $||D_{HD}D^*_{HD}|| = ||D^*_{HD}D_{HD}|| = ||D_{HD}||^2 = ||D^*_{HD}||^2$.

Proof

(1) We have

$$\begin{aligned}
(D_{HD} + T_{HD})^*h(x) &= \int_a^b h(D_{HD} + T_{HD})(x^*)\\
&= \int_a^b hD_{HD}(x^*) + \int_a^b hT_{HD}(x^*)\\
&= D^*_{HD}(h)(x^*) + D^*_{HD}(h)(x^*).
\end{aligned}$$

(2) We get

$$\begin{aligned}
(cD_{HD})^*(h)(x^*) &= \int_a^b h(cD_{HD})(x^*)\\
&= c \int_a^b hD_{HD}(x^*)\\
&= cD^*_{HD}(h)(x^*).
\end{aligned}$$

(3) Since D_{HD} and D^*_{HD} are bounded linear operators and $||D_{HD}|| = ||D^*_{HD}||$.

□

Theorem 5.11 *If $f \in D_{HD}[a, b]$, then the adjoint operator T^* is weakly compact operator.*

Theorem 5.12 *Assume* $f : [a, b] \to \mathcal{X}$ *is Henstock-Dunford integrable. The operator* $D_{HD} : \mathcal{X}^* \to HK[a, b]$ *is weakly compact if and only if the adjoint operator* $D^*_{HD} : HK^*[a, b] \to \mathcal{X}^*$ *is weakly compact operator.*

Proof According to Gantmacher's Theorem, the operator D_{HD} is a weakly compact operator if and only if its adjoint operator D^*_{HD} is also a weakly compact operator. □

5.5 Henstock-Kurzweil-Pettis Integrals

The Henstock-Kurzweil-Pettis integral is the natural generalization of the Pettis integral for a function, obtained by replacing the Lebesgue integrability for scalar functions by the Henstock-Kurzweil integrability. If $\mathcal{X}$ contains no copy of c_0, it is well known that each $\mathcal{X}$-valued strongly measurable scalarly integrable function is also Pettis integrable. In the case of the Henstock-Kurzweil integrability, a similar property is fulfilled by weakly sequentially complete Banach spaces. Gordon proved that a separable Banach space is weakly sequentially complete if and only if each $\mathcal{X}$-valued Henstock-Kurzweil scalarly integrable function is also HKP integrable.

Definition 5.8 A function $f : [0, 1] \to \mathcal{X}$ is said to be scalarly Henstock-Kurzweil integrable if for each $x^* \in \mathcal{X}^*$ the function $x^* f$ is Henstock-Kurzweil integrable. A scalarly Henstock-Kurzweil integrable function f is said to be Henstock-Kurzweil-Pettis integrable if for each $I \in [0, 1]$ there exists $w_I \in X$ such that

$$< x^*, w_I >= HK \int_I x^* f d\mu_\infty \ for\ every\ x^* \in \mathcal{X}^*.$$

We call w_I the Henstock-Kurzweil-Pettis integral of f over I and we write $w_I = (HKP) \int_I f d\mu_\infty$.

We call the additive interval function $F(I) = (HKP) \int_I f d\mu_\infty$ the HKP-primitive of f. Let $HKP([0, 1], X)$ the set of all $\mathcal{X}$-valued HKP-integrable function on [0, 1] (functions that are scalarly equivalent are identified). Henstock-Kurzweil-primitive (resp HKP-primitives) F of a function f is continuous (resp weakly continuous, i.e., $x^* F$ is continuous for every $x^* \in \mathcal{X}^*$) see [4]. The Alexiewicz norm on $HKP([0, 1], \mathcal{X})$ is

$$||f||_A = \sup_{0 \le t \le 1} \left\| (HKP) \int_0^1 f d\mu_\infty \right\|.$$

If $|x^* f| \leq K$ a.e. for some constant K and every $x^* \in \mathcal{X}^*$, then f is weakly bounded. If $f : [0, 1] \to \mathcal{X}$ is a scalarly bounded function then the image measure λ given by

$$\lambda(E) = \mu_\infty(f^{-1}(E))$$

is a weak Baire measure on $\mathcal{X}$.

Definition 5.9 Given a weak Baire measure μ_∞ on a Banach space $\mathcal{X}$, we say μ_∞ is scalarly concentrated on a subspace Γ of $\mathcal{X}$ if $x^*|_\Gamma = 0$ implies $x^* = 0$ μ_∞-a.e.

Recalling a topological space $\mathcal{X}$ is measure compact if every non-zero-σ-additive Baire measure on $\mathcal{X}$ has non-empty support. Measure-compact implies real compact but not conversely. It is well known that every compact measure is perfect. For a measure to be perfect it is necessary and sufficient that its restriction to any σ-subalgebra with a countable number of generators be compact.

Theorem 5.13 *If $f : [0, 1] \to \mathcal{X}$ is Henstock-Kurzweil-Pettis integrable, then there exists a sequence (f_n) of $\mathcal{X}$-valued step function such that for all $x^* \in \mathcal{X}^*$, $x^* f = \lim_n x^* f_n$ a.e. if and only if f is determined by a separable subspace of $\mathcal{X}$.*

Lemma 5.2 *Let $f : [0, 1] \to \mathcal{X}$ be scalarly Henstock-Kurzweil integrable. Define $T_f : \mathcal{X}^* \to HK(\mu_\infty)$ by*

$$T_f(x^*)(t) =< f(t), x^* >, \ x^* \in \mathcal{X}^*, \ t \in [0, 1]$$

then T_f is weak-pointwise continuous.*

Proof If $t \in [0, 1]$ and $\mathcal{X}_a^* \to x^*$ $(weak^*)$, then

$$\begin{aligned} T_f(x_a^*)(t) &=< f(t), x_a^* > \\ &\to< f(t), x^* > \\ &= T_f(x^*)(t). \end{aligned}$$

So, $T_f(x_a^*) = T_f(x^*)$ is pointwise. □

Proposition 5.2 *Let $f : [0, 1] \to \mathcal{X}$ be scalarly Henstock-Kurzweil integrable, then f is Henstock-Kurzweil-Pettis integrable (HKP-integrable) if and only if $T_f : \mathcal{X}^* \to HK([0, 1], \mu_\infty)$ is weak*-weakly continuous. In particular if f is Henstock-Kurzweil-Pettis integrable, then T_f is necessarily a weakly compact operator.*

Proof Let T_f is *weak**-weakly continuous. Assume $A \in \Sigma$. Then $\chi_A \in BV(\mu_\infty)$, so, the linear functional $\mathfrak{a} \in \mathcal{X}^{**}$ defined by

$$
\begin{aligned}
< x^*, \mathfrak{a} > &= HK \int_A < f(t), x^* > d\mu_\infty(t) \\
&= HK \int T_f(x^*)(t).\chi_A(t) d\mu_\infty(t) \\
&=< T_f(x^*), \chi_A >
\end{aligned}
$$

is weak* continuous, so, for $x_A \in \mathcal{X}$ with

$$< x_A, x^* >= HK \int_A < f(t), x^* > d\mu_\infty(t) \; \forall \, x^* \in \mathcal{X}^* \tag{5.1}$$

This shows f is Henstock-Kurzweil-Pettis integrable.

Conversely, assume f is Henstock-Kurzweil-Pettis integrable, then for $A \in \Sigma$ there exists $x_A \in \mathcal{X}$ with (5.1). So, from the representation theorem and convergence theorem we have

$$< T_f(x^*), \chi_A >= HK \int_A < f(t), x^* > d\mu_\infty(t) =< x_A, x^* >$$

is weak*-continuous function of x. Hence $< T_f(x^*), g >$ is weak* continuous function of x^* for any step function g. Again, using the boundedness of f, we get $< T_f(x^*), g >$ is weak*-continuous function of x^* for any $g \in BV(\mu_\infty)$. □

Corollary 5.1 *Let $f : [0, 1] \to \mathcal{X}$ be scalarly Henstock-Kurzweil integrable, then f is HKP-integrable if and only if $T_f : B(\mathcal{X}^*) \to HK([0, 1], \mu_\infty)$ is weak*-weakly continuous.*

From the Proposition 5.2 and the Corollary 5.1, we can deduce the following corollary.

Corollary 5.2 *A scalarly Henstock-Kurzweil integrable function $f : [0, 1] \to \mathcal{X}$ determined by a subspace Y of $\mathcal{X}$ if and only if $T : \mathcal{X} \to HK(\mu_\infty)$ defined by $x^* \to x^* f$ is $\sigma(\mathcal{X}^*, Y)$-weakly continuous.*

Recalling the action of T_f on weak* neighborhoods in $\mathcal{X}^*$ as follows:

Definition 5.10 Let Y be a finite set in $\mathcal{X}$, and $\epsilon > 0$, let

$$O(Y, \epsilon) = \left\{ x^* \in \mathcal{X}^* : ||x^*|| \leq 1 \text{ and } x^*(x) \leq \epsilon \text{ for every } x \in Y \right\}.$$

Proposition 5.3 *If f is scalarly Henstock-Kurzweil integrable, then for all $Y \subseteq \mathcal{X}$, $\epsilon > 0$ the set $T_f\,(O(Y, \epsilon))$ is closed and convex in $HK(\mu_\infty)$.*

Proof The proof is of similar argument of the [17, Lemma 2]. □

Proposition 5.4 *If f is scalarly Henstock-Kurzweil integrable. Then the following are equivalent:*

1. *$f : [0, 1] \to \mathcal{X}$ is Henstock-Kurzweil-Pettis integrable.*
2. *T_f is a weakly compact operator and*

$$\mathbb{Z} = \bigcap \{T\left((O(Y, \epsilon)\right) : Y \subset \mathcal{X},\ Y\ is\ finite,\ \epsilon > 0\} = \{0\}.$$

Proof The proof is of similar argument of the [17, Proposition 3]. □

5.5.1 Henstock-Kurzweil-Pettis Integral Property

Here we give several conditions of Henstock-Kurzweil-Pettis integrability in $\mathcal{X}$ in the language of weak Baire measure and compact measure. $\mathcal{X}$ has the μ_∞-HKP integral property if each $\mathcal{X}$-valued, scalarly μ_∞-measurable function is μ_∞-HKP integrable. If such a property holds true for all complete measure, then $\mathcal{X}$ is said to have the HKP-integral property. A Banach space $\mathcal{X}$ is said to have Henstock-Kurzweil-Pettis Integral property (HKPIP) if for every $\mathcal{X}$-valued weakly bounded function f defined on an arbitrary finite measure space is Henstock-Kurzweil-Pettis integrable.

Theorem 5.14 *If $\mathcal{X}$ is μ_∞-measure compact, then $\mathcal{X}$ has the μ_∞-HKPIP. If $\mathcal{X}$ is measure compact, then X has HKPIP.*

Proof Suppose $\mathcal{X}$ is μ_∞-measure compact. Let $f : [0, 1] \to \mathcal{X}$ is bounded and scalarly measurable, then f is weakly equivalent to Henstock-Kurzweil measurable function $g : [0, 1] \to \mathcal{X}$. Since g is bounded and measurable, hence Henstock-Kurzweil integrable. If x_A is Henstock-Kurzweil integral, i.e., $x_A = HK \int_A g d\mu_\infty$, then for every $x^* \in \mathcal{X}^*$ the function $x^* f = x^* g$ a.e. So,

$$\begin{aligned} x_A x^* &= HK \int_A g x^* d\mu_\infty \\ &= HK \int_A f x^* d\mu_\infty. \end{aligned}$$

Hence x_A is Henstock-Kurzweil-Pettis integral $(HKP) \int_A f d\mu_\infty$. □

Theorem 5.15 *If the condition of Mazur in $\mathcal{X}$, then HKPIP does not hold.*

Proof The condition of Mazur in $\mathcal{X}$ implies Pettis Integral Property in $\mathcal{X}$. If $\mathcal{X}$ has the property of Mazur implies not naturally HKPIP in $\mathcal{X}$. Ye Guoju et Al. [14] presented that Pettis integrals are subset of Henstock-Kurzweil-Pettis integral Mazur property in $\mathcal{X}$ (particularly separable or weakly compactly generated) implies HKPIP if and only if $f : [0, 1] \to \mathcal{X}$ is scalarly Henstock-Kurzweil integrable and each infinite subset of $\{x^* f : ||x^*|| \leq 1\}$ contains an equi-Henstock-Kurzweil integrable sequences. □

5.6 Henstock-Kurzweil-Pettis Integrals with Separable Range

In this section we discuss about separable ranges of HKP-integrable function. Recalling the following theorem:

Theorem 5.16 *Suppose $f : [0, 1] \to \mathcal{X}$ be a function. If there exists a sequence of Henstock-Kurzweil-Pettis integrable functions $f_n : [0, 1] \to \mathcal{X}$ such that*

1. *The set $\{x^* f_n : ||x^*|| \leq 1, \ n \in \mathbb{N}\}$ is uniformly μ_∞-integrable.*
2. *$\lim_n x^* f_n = x^* f$ in μ_∞-measurable for each $x^* \in \mathcal{X}^*$.*

Then clearly f is Henstock-Kurzweil-Pettis μ_∞-integrable and

$$\lim_n HK \int_E f_n d\mu_\infty = HK \int_E f d\mu_\infty \ E \subseteq [0, 1]$$

weakly in X for each $E \in \Sigma$.

Theorem 5.17 *If $f \in HKP([0, 1], \mathcal{X})$, then the following are equivalent:*

1. *$\{x^* f : x^* \in B(\mathcal{X}^*)\}$ is separable subset of $HK([0, 1])$.*
2. *There exists a sequence (f_n) of $\mathcal{X}$-valued step function, such that for each $x^* \in X^*$ the sequence $(x^* f_n)_{n \in \mathbb{N}}$ converges weakly to $x^* f$ in $HK([0, 1])$.*
3. *$\mu_{\infty_f}(\Sigma)$ is a separable subset of $\mathcal{X}$.*

Proof For (1) $\implies$ (2) : Let $\{x^* f : x^* \in B(\mathcal{X}^*)\}$ is separable. This means there exists a sequence (x_n^*) in $B(X^*)$ such that $\{x_n^* f : n \in \mathbb{N}\}$ is dense in $\{x^* f : x^* \in B(\mathcal{X}^*)\}$. Now $f \in HKP([0, 1], \mathcal{X})$, then the topology of convergence in measure and the weak topology of $HK[0, 1]$ coincide on the set $\{x^* f : x^* \in B(\mathcal{X}^*)\}$, and there exists a step function $f_n : [0, 1] \to \mathcal{X}$ such that $x^* f_n \to x^* f$ weakly in $HK[0, 1]$ for every $x^* \in \mathcal{X}^*$.

For (2) $\implies$ (3) : By the conditions of (2) for each $E \in \Sigma$ the sequence $\mu_{\infty_{fn}}(E)$ is weakly convergent to $\mu_{\infty_f}(E)$. So, $\mu_{\infty_f}(\Sigma)$ contained in the weak closure of the set $\bigcup_{n=1}^{\infty} \mu_{\infty_{fn}}(\Sigma)$ as separable due because the ranges of $\mu_{\infty_{fn}}$ are finite dimensional.

For (3) $\implies$ (1) : We use contradiction method. If possible assume $\{x^* f \ : x^* \in B(\mathcal{X}^*)\}$ is non-separable and $x_1^* \in U(\mathcal{X}^*)$ where

$$U(\mathcal{X}^*) = \{(x, s) : | < x^*, f(x) > | \geq s||x^*||\} \ for \ x^* \in \mathcal{X}^*$$

and $h \in BV(\mu_\infty)$ such that $< h, x_1^* f >= 1$. Let for an ordinal $b < x_1$, the family $\{(x_a^*, h_a) : a < b\}$ satisfies the following:

a. $x_a^* \in U(\mathcal{X}^*)$,
b. $h_a \in BV(\mu_\infty)$,
c. $x_c^* f \in \overline{\lim}\{x_a^* f : a < c\}$ for each $c < b$,
d.

$$< h_c, x_a^* f >= \begin{cases} 1 \text{ if } a = c < b \\ 0 \text{ if } a < c < b. \end{cases}$$

As $\{x^* f : x^* \in B(\mathcal{X}^*)\}$ is non-separable implies $x_b^* \in U(\mathcal{X}^*)$ such that $x_b^* f \notin \overline{\lim}\{x_a^* f : a < b\}$. Now using Hahn-Banach Theorem $h_b \in BV(\mu_\infty)$ such that $< h_b, x_b^* f >= 1$ and $< h_b, x_a^* f >= 1$ for all $a < b$. From the net $\{(x_a^*, h_a) : a < x_1\}$ satisfying $(a) - (d)$ for all a, b, c less than x_1 implies $||T_f^*(h_b) - T_f^*(h_a)|| \geq 1$ whenever $a < b$. So, $T_f^*(BV(\mu_\infty))$ is non-separable on X^{**} but $\{\chi_E : E \in \Sigma\}$ is norm dense in $BV(\mu_\infty)$ and so $\mu_\infty(\Sigma)$ is norm dense in $T_f^*(BV(\mu_\infty))$, a contradiction. Hence $\mu_{\infty_f}(\Sigma)$ is separable. □

By amalgamating Theorem 5.16 and Theorem 5.17, we can find out the following conclusion.

Theorem 5.18 *Let $f : [0, 1] \to \mathcal{X}$ be a function, then $f \in HKP([0, 1], \mathcal{X})$ where μ_∞ is a perfect measure and $\mu_{\infty_f}(\Sigma)$ is a separable set if and only if there exists a sequence (f_n) of $\mathcal{X}$-valued step function such that*

1. *The family $\{x^* f_n : n \in \mathbb{N},\ x^* \in B(\mathcal{X}^*)\}$ is uniformly integrable.*
2. *For each $x^* \in \mathcal{X}^*$ we have $\lim_n x^* f_n = x^* f$ a.e.*

With the analogous way, we can construct the following Lemma from Talagrand [24, 5-1-2], which was used in [20] by G. Plebanek in his work.

Lemma 5.3 *Let $([0, 1], \Sigma, \mu_\infty)$ be a measure space and C be an absolutely convex set of measurable function which is τ_m-compact and bounded in $BV[0, 1]$. Then the following are equivalent:*

1. *$\mu_\infty : C \to \mathbb{R}$, $\mu_\infty(g) = HK \int_0^1 g d\mu_\infty$ is τ_m-continuous.*
2. *There exists a countable subset $D \subset [0, 1]$ such that $(HK) \int_0^1 g d\mu_\infty = 0$ whenever $g \in C$ and $g_{|D} = 0$.*

Theorem 5.19 *If $f : [0, 1] \to \mathcal{X}$ is scalarly Henstock-Kurzweil integrable function, then $f \in HKP([0, 1], \mathcal{X})$ and $\mu_{\infty_f}(\Sigma)$ is separable if and only if the measure $f(\mu_\infty)$ is scalarly concentrated on a separable subspace of $\mathcal{X}$.*

Proof Let $f : [0, 1] \to \mathcal{X}$ is Henstock-Kurzweil-Pettis integrable function and $\mu_{\infty_f}(\Sigma)$ is separable. Let Γ be separable subspace of $\mu_{\infty_f}(\Sigma)$. From the [8, Theorem 3] we get $x^* \in \Gamma^\perp$ and $x^* f = 0$ a.e. so, $x^* = 0$ $f(\mu_\infty)$-a.e.

Conversely, let $\nu = f(\mu_\infty)$ and $\mathbb{G}$ the closure of a countable set of X such that ν is scalarly concentrated on $\mathbb{G}$. Now $x^*_{|\mathbb{G}} = 0$ implies $x^* = 0$ ν-a.e. From the Lemma 5.3 $x^* \rightarrow HK \int_B x^* d\nu$ is τ_m-continuous for every $B \in Ba(\mathcal{X})$. Hence, $\mathcal{X}^* \rightarrow HK(\nu)$ is $\tau_m - (weak)$ continuous, where τ_m is the topology of pointwise convergence and $i_{\mathcal{X}} : \mathcal{X} \rightarrow \mathcal{X}$ is Henstock-Kurzweil-Pettis integrable with respect to ν. If $x_0 = HK \int_B i_{\mathcal{X}} d\nu$, then $x^*_{|\mathbb{G}} = 0$ implies $x^* x_0 = 0$, hence $x_0 \in \mathbb{G}$. This shows f is Henstock-Kurzweil-Pettis integrable and the range $\mu_{\infty_f}(\Sigma)$ is separable on $\mathbb{G}$. □

We begin with a simple characterization of indefinite Henstock-Kurzweil-Pettis integrals that have separable ranges.

Lemma 5.4 *Suppose $\mathcal{X}$ is separable Banach spaces, $f : [0, 1] \rightarrow \mathcal{X}$ is Henstock-Kurzweil-Pettis integrable on $[0, 1]$ and let F be the indefinite integral of f, then the following are equivalent:*

1. *F is weak*-continuous on $[0, 1]$.*
2. *The set $\left\{\int_I f : I \subset [0, 1]\right\}$ is separable.*

Proof For (1) $\Longrightarrow$ (2) : Let F is $weak^*$-continuous on $[0, 1]$, then $F([0, 1])$ is weak*- compact. So, $\left\{\int_I f : I \subset [0, 1]\right\}$ is weak*-compact then the weak*-topology of $\mathcal{X}^*$ restricted to $\left\{\int_I f : I \subset [0, 1]\right\}$ is metrizable. Hence, $\left\{\int_I f : I \subset [0, 1]\right\}$ is separable.

For (2) $\Longrightarrow$ (1) : Let (f_n) be a sequence of step function on $[0, 1]$ and F_n be the indefinite integral of f_n, then

$$\lim_n ||x^* F - x^* F_n||_n = 0.$$

Now since F_n is weak*-continuous on $[0, 1]$. This shows F is weak*-continuous on $[0, 1]$. □

Theorem 5.20 *Let $f : [0, 1] \rightarrow \mathcal{X}$ be scalarly Henstock-Kurzweil integrable function and $\mathcal{X}$ is a Mazur space, then following are held:*

1. *If $f : [0, 1] \rightarrow \mathcal{X}$ is Henstock-Kurzweil-Pettis integrable on $[0, 1]$, then the indefinite integral of f is weak*-continuous on $[0, 1]$.*
2. *If $f : [0, 1] \rightarrow \mathcal{X}$ is Henstock-Kurzweil-Pettis integrable on $[0, 1]$, then the set $\left\{\int_I f : I \subset [0, 1]\right\}$ is separable.*

Proof For (1) : Let F be the indefinite integral of f. Since $x^* F$ is weak*-continuous on $[0, 1]$. Due to the Mazur property of $\mathcal{X}$, we can find F is weak*-continuous on $[0, 1]$.

For (2) : it is the direct implication of the Lemma 5.4. □

Lemma 5.5 *Let $f : [0, 1] \rightarrow \mathcal{X}$ be Henstock-Kurzweil-Pettis integrable with $\mathcal{X}$ is separable and $T : \mathcal{X}^* \rightarrow HK[0, 1]$ be a compact, then the range $R_{\mu_\infty}\{F(I)\}$ is separable if and only if the set $\{x^* f : ||x^*|| \leq 1\}$ is separable subset of $HK[0, 1]$.*

Proof Let $T : \mathcal{X}^* \to HK[0,1]$ defined by $T(x^*) = x^* f$ be compact. Then $T^* : BV[0,1] \to \mathcal{X}^{**}$ is such that

$$\begin{aligned} < x^*, T^*_{\chi_I} > &=< T(x^*), \chi_I > \\ &= HK \int_I x^* f d\mu_\infty \\ &=< x^*, (HKP) \int_I f d\mu_\infty > \\ &=< x^*, F(I) > . \end{aligned}$$

That is $F(I) = T^*_{\chi_I}$. Then $F(I)$ is compact by Schauder Theorem. Let $\mathcal{X} = \bigcup_n B(0,n) T(B(0,n))$ be separable and $\bigcup_n T(B(0,n)) = R_{\mu_\infty}(F)$ is separable, since the union of separable spaces. □

Theorem 5.21 *If $\mathcal{X}$ is separable Banach space and let $f : [0,1] \to \mathcal{X}$ be Henstock-Kurzweil-Pettis integrable function. Assume $F : I \to \mathcal{X}$ be its primitive, then the following are equivalent:*

1. *The family $\{x^* f_n : n \in \mathbb{N},\ x^* \in B(\mathcal{X}^*)\}$ is uniformly integrable.*
2. *For each $x^* \in \mathcal{X}^*$ we have $\lim_n x^* f_n = x^* f$ a.e.*
3. *$R_{\mu_\infty}(F)$ is separable subset of $\mathcal{X}$, where $R_{\mu_\infty}(F)$ is the range spaces.*

Proof For (1) $\implies$ (2) is (1) $\to$ (2) of the Theorem 5.17.

For (2) $\implies$ (3) : from the condition (2) for each $E \in \Sigma$ the sequence $\{HK \int_E f_n d\mu_\infty : n \in \mathbb{N}\}$ is weakly convergent to $HK \int_E f d\mu_\infty$. Hence $R_{\mu_\infty}(F)$ is contained in the weak closure of the set $\bigcup_n F_n(\Sigma)$ where F_n is the indefinite Henstock-Kurzweil-Pettis integral of f_n. Each set $F_n(\Sigma)$ is of finite dimension, so the union is weakly separable. Now by using Mazur Theorem $R_{\mu_\infty}(F)$ is separable as the weak and norm separability coincide in Banach space.

For (3) $\implies$ (1) is similar as (3) $\implies$ (1) of Theorem 5.17. □

Lemma 5.6 *Let Y be a subspace of $\mathcal{X}$ and assume Y contains no isomorphic copy of c_0. If Y is WCG (has Mazur's property), then $f : [0,1] \to \mathcal{X}$ a bounded scalarly Henstock-Kurzweil integral determined by Y is Henstock-Kurzweil-Pettis integrable.*

Proof Let Y be WCG and f be scalarly Henstock-Kurzweil integrable. Then there exists a countable partition P_i of $[0,1]$ into measurable sets such that for each E in P_i, the function $f.\chi_E$ is weakly bounded. Since f is Pettis integrable and hence Henstock-Kurzweil-Pettis integral for all E in P_i. This shows for each

$F \in \Sigma$, $HK \int_{F\cap E} f d\mu_\infty = HKP \int_{F\cap E} f d\mu_\infty \in X$. Now, for $x^* \in X^*$ and $F \in \Sigma$,

$$\sum_{E\in P_i} \left| x^* \left((HKP) \int_{F\cap E} f d\mu_\infty \right) \right| = \sum_{E\in P_i} |HK \int_{F\cap E} x^* f d\mu_\infty|$$
$$\leq \sum_{E\in P_i} HK \int_{F\cap E} |x^* f| d\mu_\infty$$
$$= HK \int_F |x^* f| d\mu_\infty < \infty.$$

Now by Bessaga-Pelezynski characterization theorem, the series $\sum_{E\in P_i} (HKP) \int_{F\cap E} f d\mu_\infty$ is an unconditionally norm convergent series for all F in Σ. That is, $\sum_{E\in P_i} (HKP) \int_{F\cap E} f d\mu_\infty = (HKP) \int_F f d\mu_\infty$. □

Proposition 5.5 *Let $f : [0, 1] \to \mathcal{X}$ be scalarly Henstock-Kurzweil integrable and $T_f : \mathcal{X}^* \to HK([0, 1])$ defined by $x^* \to x^* f$ is weakly compact. If f is determined by a subspace having Mazur's property, then f is Henstock-Kurzweil-Pettis integrable.*

Proof Let f be determined by $Y \subset \mathcal{X}$ having Mazur's property. This shows each *weak**-continuous functional on Y^* is in Y. Let (x_n) be dense in Y. For each n, choose x_n^* in $O\left(\{x_j\}_{j=1}^n, \epsilon\right)$ so that $x_n^* f = g$ a.e. for $g \in \bigcap_{(Y,\epsilon)} T(O(Y, \epsilon))$ if possible. We need to claim $g = 0$ a.e. Let x^* be weak* cluster of $(x_n^*)_n$, then $g = x^* f$ a.e. From Theorem 5.16 $\lim_n x_n^* f = x^* f = g$ a.e. The Mazur's property in Y gives $x^* f = g = 0$. □

References

1. Alexiewicz, A.: Linear functionals on Denjoy-integrable functions. Collectanea Mathematica **1**, 289–293 (1948)
2. Aqzzouz, B., Elbour, A.: Characterization of the order weak compactness of semi-compact operators. J. Math. Anal. Appl. **355**, 541–547 (2009)
3. Barr, M.: Duality of Banach spaces. Cahiers de topologie et géométrie différentielle catégoriques **17**(1), 15–32 (1976)
4. Bongiorno, B., Di. Piazza, L., Musiał', K.: Approximation of Banach space valued Non-absolutely integrable functions by step functions. Glasgow Math. J. **50**, 583–593 (2008)
5. Bruckner, A.M., Bruckner, J.B., Thomson, B.S.: Real Analysis, 2nd edn. Classical Real Analysis.com (2008)
6. Candeloro, D., Di Piazza, L., Musial, K., Sambucini, A.R.: Some new results on integration for multifunctions. Ricerche di Matematica **67**, 361–372 (2018)
7. Candeloro, D., Di Piazza, L., Musial, K., Sambucini, A.R.: Integration of multifunctions with closed convex valued in arbitrary Banach spaces. J. Convex Anal. **27**(4), 1233–1246 (2020)

8. Di Piazza, L., Musiał, K.: Characterizations of Kurzweil-Henstock-Pettis integrable functions. Studia Mathematica **176**(2), 159–176 (2006)
9. Di Piazza, L., Marraffa, V., Musial, K.: Variational Henstock integrability of Banach space valued functions. Mathematica Bohemica **141**, 287–296 (2016)
10. Di Piazza, L., Musial, K.: Set valued Kurzweil-Henstock Pettis integral. Set-Valued Anal. **13**, 167–169 (2005)
11. Gill, T.L., Zachary, W.W.: Functional Analysis and Feynman Operator Calculus. Springer, Cham (2016)
12. Gordon, R.A.: The Denjoy extension of the Bochner, Pettis, and Dunford integrals. Studia Mathematica **92**(1), 73–91 (1989)
13. Gordon, R.: The Integrals of Lebesgue, Denjoy, Perron, and Henstock. American Mathematical Society, Providence (1994)
14. Guoju, Y., Tianguing, A.N.: On Henstock-Dunford and Henstock-Pettis integral. Int. J. Math. Math. Sci. **25**(7), 467–478 (2001)
15. Hazarika, B., Kalita, H.: Henstock-Orlicz spaces and its dense spaces. Asian Eur. J. Math. **14**(7), 1–17 (2021)
16. Kalita, H., Hazarika, B.: Countable additivity of Henstock-Dunford integrable functions and Orlicz space. Anal. Math. Phys. **11**(2), 1–13 (2021)
17. Kluvánek, I., Knowles, G.: Vector Measures and Control Systems. Mathematics Studies. North Holland Publishing Company, Amsterdam (1975)
18. Lindenstrauss, J., Tzafriri, L.: Classical Banach Spaces II, Function Spaces. A Series of Modern Surveys in Mathematics. Springer, Berlin-Heidelberg-New York (1979)
19. Perez-Becerra, T., Sanchez-Perales, S., Oliveros-Oliveros, J.J.: The HK-Sobolev space and applications to one-dimensional boundary value problems. J. King Saud University-Science (2020). https://doi.org/10.1016/j.jksus.2020.06
20. Plebanek, G.: On Pettis integrals with separable range. Colloquium Mathematicae **64**(1), 71–78 (1993)
21. Rao, M.M., Ren, Z.: The Theory of Orlicz Spaces. Marcel Dekker, New Work (1991)
22. Schwabik, S., Guoju, Y.: Topics in Banach Spaces Integration, Series in Real Analysis, vol 10. World Scientific, Singapore (2005)
23. Solikhin, S., Hariyanto, S., Sumanto, Y.D., Aziz, A.: Weakly compact linear operators on space of Dunford integral function. J. Phys. Conf. Ser. **1524**, 012045 (2020)
24. Talagrand, M.: Pettis integral and measure theory. Mem. Am. Math. Soc. **51**, 2–25 (1984)
25. Yamasaki, Y.: Measures on Infinite-Dimensional Spaces. Series in Pure Mathematics, vol. 5 (1985)

Chapter 6
Henstock-Orlicz Spaces and Denseness of C_0^∞

In this chapter we define the Henstock-Orlicz class for bounded measurable functions with compact support and discuss some fundamental results of this class. Further, we construct H-Orlicz spaces. Several fundamental properties of H-Orlicz space have been discussed. Lastly we prove C_0^∞ is a dense subset of H-Orlicz space.

6.1 Structure of H-Orlicz Classes

We assume an abstract measure space $(\mathcal{B}, \Sigma, \mu_\infty)$, where $\mathcal{B}$ is some set of points and Σ is an σ-algebra of its subsets on which a σ-additive function $\mu_\infty : \Sigma \to \mathbb{R}^+$ is given. In this section we define the Henstock-Orlicz class for bounded measurable functions with compact support and discuss some fundamental results of this class. We assume an abstract measure space $(\mathcal{B}, \Sigma, \mu_\infty)$, where $\mathcal{B}$ is some set of points and Σ is an σ-algebra of its subsets on which a σ-additive function $\mu_\infty : \Sigma \to \mathbb{R}^+$ is given. Now onward we assume $H(\mu_\infty)$ is the Henstock-Kurzweil (in short Henstock) integrable function space.

Theorem 6.1 *Let $I_0 \subset I$. If $f : I_0 \subseteq \mathcal{B} \to \mathbb{R}$ is a measurable function with gauge function $\delta : I_0 \subseteq \mathcal{B} \to \mathbb{R}$, then $\theta(f) : \mathcal{B} \to \mathbb{R}^+$ is Henstock integrable.*

Proof Let $f : I_0 \subseteq \mathcal{B} \to \mathbb{R}^+$ be measurable function, then for all $\epsilon > 0$ there exists a $\delta : I_0 \to \mathbb{R}$ such that

$$\left| \sum_{(I,w)\in\pi} \sum_{(I',w')\in\pi'} [f(w) - f(w')]\mu_\infty(I \cap I') \right| < \epsilon$$

for all partitions π and π' of I_0 finer than δ.

B. Hazarika, H. Kalita, *Theory of Henstock–Orlicz Spaces*, Mathematical Marvels: Texts and Monographs in the Spirit of CR Rao,
https://doi.org/10.1007/978-981-96-9548-5_6

As θ is Young function. So, $\theta(x) \to \infty$ as $x \to \infty$. Our claim is $\theta(f)$ is Henstock integrable.

Since Young function by definition is an extended real Borel function. So, $\theta(f)$ is measurable. If π and $\pi^{'}$ are both partitions of the same interval of I_0, then for any subinterval I of I_0 we can write

$$\mu_\infty(I) = \sum_{(I^{'},w^{'})\in\pi^{'}} \mu_\infty(I \cap I^{'}).$$

So,

$$\sum_{(I,w)\in\pi} \theta(f)(w)\mu_\infty(I) = \sum_{(I,w)\in\pi} \sum_{(I^{'},w^{'})\in\pi^{'}} \theta(f)(w)\mu_\infty(I \cap I^{'})$$

$$\text{i.e.} \left| \sum_{(I,w)\in\pi} \theta(f)(w)\mu(I) - \sum_{(I^{'},w^{'})\in\pi^{'}} \theta(f)(w)\mu_\infty(I) \right| < \epsilon.$$

Thus, $\theta(f)$ is Henstock integrable. □

Definition 6.1 Let $\mathcal{H}^{-\theta}(\mu_\infty)$ be the set of all $f : \mathcal{B} \to \mathbb{R}$ bounded measurable with compact support for $\Sigma \subset \mathcal{B}$ such that $\int_{\mathcal{B}} \theta(|f|)d\mu$ is Henstock integrable.

That is, $\mathcal{H}^{-\theta}(\mu_\infty) = \{f$ is bounded measurable with compact support : $\int_{\mathcal{B}} \theta(|f|)d\mu_\infty \in H(\mu_\infty)\}$.

Theorem 6.2 *The space $\mathcal{H}^{-\theta}(\mu_\infty)$ is absolutely convex. That is, if $f, g \in \mathcal{H}^{-\theta}(\mu_\infty)$ and α, β are scalars such that $|\alpha|+|\beta| \leq 1$, then $\alpha f+\beta g \in \mathcal{H}^{-\theta}(\mu_\infty)$. Also $h \in \mathcal{H}^{-\theta}(\mu_\infty)$, $|f| \leq |h|$, f is Henstock integrable, then $f \in \mathcal{H}^{-\theta}(\mu_\infty)$.*

Proof Let $f, g \in \mathcal{H}^{-\theta}(\mu_\infty)$. Then by the monotonicity and convexity of θ, we get $0 < \gamma = |\alpha| + |\beta| \leq 1$. Now,

$$\begin{aligned}
\theta(|\alpha f + \beta g|) &\leq \theta(|\alpha||f| + |\beta||g|) \\
&\leq \gamma\theta\left(\frac{|\alpha|}{\gamma}|f| + \frac{|\beta}{\gamma}|g|\right) \\
&\leq |\alpha|\theta(|f|) + |\beta|\theta(|g|).
\end{aligned}$$

The right-hand side is Henstock integrable. So, $\alpha f + \beta g \in \mathcal{H}^{-\theta}(\mu_\infty)$.

For second part $|f| \leq |h|$. If $f \neq h$, then for all $x \in \mathcal{B}$, there exists $\epsilon > 0$ such that $|h - \epsilon| \leq |f| \leq |h|$.

As $h \in \mathcal{H}^{-\theta}(\mu_\infty)$ gives $\theta(|h|)$ is Henstock integrable. Also, $\theta(|h - \epsilon|)$ is Henstock integrable.

So, by [9, Lemma 9], $f \in \mathcal{H}^{-\theta}(\mu_\infty)$. □

Theorem 6.3 *The space $\mathcal{H}^{-\theta}(\mu_\infty)$ is a linear space if and only if θ satisfies Δ_2-condition.*

Proof For linearity, it is sufficient to verify that for each $f \in \mathcal{H}^{-\theta}(\mu_\infty), 2f \in \mathcal{H}^{-\theta}(\mu_\infty)$ since then $nf \in \mathcal{H}^{-\theta}(\mu_\infty)$ for any integer n. Since $h \in \mathcal{H}^{-\theta}(\mu_\infty), |f| \leq |h|, f$ measurable, implies $f \in \mathcal{H}^{-\theta}(\mu_\infty)$ hence for each $\alpha > 0,\ \alpha f \in \mathcal{H}^{-\theta}(\mu_\infty)$.

Then yields, $\alpha f + \beta g = \gamma\left(\frac{\alpha}{\gamma} f + \frac{\beta}{\gamma} g\right) \in \mathcal{H}^{-\theta}(\mu_\infty),\ \gamma = |\alpha| + |\beta| > 0$ for any $f, g \in \mathcal{H}^{-\theta}(\mu_\infty)$.

If θ is Δ_2-regular, then $\mu_\infty(\mathcal{B}) = +\infty$ and $\theta(2|f|) \leq k\theta(|f|),\ c > 0$ implies $2f \in \mathcal{H}^{-\theta}(\mu_\infty)$.

If $\mu_\infty(\mathcal{B}) < \infty$, then $\theta(2x) \leq c\theta(x)$ for $x \geq x_0 \geq 0$.

Let

$$f_1 = \begin{cases} f, & \text{if } |f| \leq x_0, \\ 0, & \text{otherwise} \end{cases}$$

We set $f_2 = f - f_1$ that is $f = f_1 + f_2$ and

$$\begin{aligned} \theta(2|f|) &= \theta(2|f_1 + f_2|) \\ &\leq \theta(2|f_1|) + \theta(2|f_2|) \text{ (by convexity of } \theta) \\ &\leq \theta(2|f_1|) + k\theta(2|f_2|) \text{ (by } \Delta_2 - \text{condition)}. \end{aligned}$$

Therefore $\int_{\mathcal{B}} \theta(2|f|) d\mu_\infty$ is Henstock integrable as right side is Henstock integrable.

So, $2f \in \mathcal{H}^{-\theta}(\mu_\infty)$. That is $nf \in \mathcal{H}^{-\theta}(\mu_\infty)$ for any integer n.

Therefore $\mathcal{H}^{-\theta}(\mu_\infty)$ is linear when $\theta \in \Delta_2$.

For converse, let $E \in \Sigma$ be a set of positive measure on which μ_∞ is diffuse and that θ does not belong in Δ_2.

We construct $f \in \mathcal{H}^{-\theta}(\mu_\infty)$ such that $2f \notin \mathcal{H}^{-\theta}(\mu_\infty)$.

If $0 < \alpha < \mu_\infty(E) < \infty$. Then by hypothesis on μ_∞, there is $F \subset E,\ F \in \Sigma$ with $\mu(F) = \alpha < \infty$.

We construct a function supported by F to satisfy our assumption. Let $\theta(\mathbb{R}) \subset \mathbb{R}^+$. Since θ does not satisfy Δ_2-condition.

There exists a sequence $x_j \geq j$ such that $\theta(2x_j) > n\theta(x_j),\ j \geq 1$.

Let n_0 be an integer such that

$$\sum_{n \geq n_0} \frac{1}{n^2} < \infty$$

and $\theta(x_n) \geq 1$ for all $n \geq n_0$.

This is possible by diffuseness of μ_∞ on F there is a measurable $F_0 \subset F$ such that

$$\mu(F_0) = \sum_{n \geq n_0} \frac{1}{n^2} < \infty.$$

Similarly we can find a set $D_1 \in \Sigma$, $D_1 \subset F_0$ such that $\mu_\infty(D_1) = \frac{1}{n^2}$.

Since $\mu_\infty(F_0 - D) > 0$, we can again find $D_2 \in \Sigma$ and $D_2 \subset F_0 - D_1$ such that $\mu_\infty(D_2) = \frac{1}{(n_0+1)^2}$.

Then for disjoint sets $D_n \in \Sigma,\ \mu_\infty(D_n) = \frac{1}{(n_0+n-1)^2},\ n \geq 1$.

Let $F_k \subset D_k,\ F_k \in \Sigma$, be chosen such that $\mu_\infty(F_k) = \frac{\mu_\infty(D_k)}{\theta(x_n)}$.

Let

$$\sum_{j=1}^{n} \theta(x_j)\mu_\infty(F_j) = \int_{I_0} f.$$

Then $f \in \mathcal{H}^{-\theta}(\mu_\infty)$ (Remark 5 of [9]). However

$$\begin{aligned}
\int_{\mathcal{B}} \theta(2f)d\mu_\infty &= \sum_{j=1}^{n} \theta(2x_j)\mu_\infty(F_j) \\
&\geq \sum_{n \geq n_0} j\theta(x_j)\mu_\infty(F_j) \\
&= \sum_{n \geq n_0} \frac{1}{j} = +\infty.
\end{aligned}$$

So, $2f \notin \mathcal{H}^{-\theta}(\mu_\infty)$.

So, our conclusion $\mathcal{H}^{-\theta}(\mu_\infty)$ is a class of scalar function, and it is linear if and only it is closed under multiplication by positive number. □

Example 6.1 Consider $\mathcal{B} = \{[i-1, i]\ : i \in \mathbb{N}\}$. Let Σ be power set of $\mathcal{B}$ and μ_∞ be positive measure with the ith interval has measure 1, and let $\theta(x) = e^{x^2} - 1$. Then $\theta(x)$ is N-function such that $\theta \notin \Delta_2$, we have $\mathcal{H}^{-\theta}(\mu_\infty)$ is linear space.

Proof By definition $\theta(x)$ is N-function and it does not satisfy Δ_2-condition. We will see $\mathcal{H}^{-\theta}(\mu_\infty)$ is linear space.

If $f \in \mathcal{H}^{-\theta}(\mu_\infty)$, then $\int_{\mathcal{B}} \theta(f)d\mu_\infty = \sum_{n=1}^{\infty}(e^{|f|^2} - 1) < \infty$.

So, $\int_{\mathcal{B}} \theta(f)d\mu_\infty$ is bounded and Henstock integrable.

Let $c > 0$ be bound. So $(e^{|f|^2} - 1) \leq c$ implies $e^{|f|^2} \leq 1 + c$.

Therefore $\int_{\mathcal{B}}(\theta(2f)d\mu_\infty = \sum_{n=1}^{\infty}\left(e^{4|f|^2} - 1\right)$.

That is

$$\int_{\mathcal{B}} \theta(2f)d\mu_\infty \leq \sum_{n=1}^{\infty}(\exp(|f(x)|^2) - 1)(c+2)((c+1)^2+1)$$
$$\leq [(c+2)(c+1)^2+1]\int_{\mathcal{B}} \theta(f)d\mu_\infty.$$

Right side is Henstock integrable. So $2f \in \mathcal{H}^{-\theta}(\mu_\infty)$ and the space is linear. □

Proposition 6.1 *Let* $(\mathcal{B}, \Sigma, \mu_\infty)$ *be finite measure space. Then* $HK(\mu_\infty) = \bigcup\{\mathcal{H}^{-\theta}(\mu_\infty) : \theta$ *ranges over all* $N-$*function.*$\}$

Proof Since θ is convex. $\theta(x) \geq ax + b$ for some constant a, b. For each $f \in \mathcal{H}^{-\theta}(\mu_\infty)$, we have $\int_{\mathcal{B}}(a|f| + b)d\mu_\infty \leq \int_{\mathcal{B}}\theta(|f|)d\mu_\infty$.

Since $\int_{\mathcal{B}}\theta(|f|)d\mu_\infty$ is Henstock integrable. So f is bounded measurable function. Using Lusin's theorem and the continuity of f (Exercise 13 of [4]), we have f as Henstock-Kurzweil integrable, so $f \in HK(\mu_\infty)$. Since this holds for all such convex function. Therefore $\bigcup \mathcal{H}^{-\theta}(\mu_\infty) \subset HK(\mu_\infty)$.

If $f \in HK(\mu_\infty)$, and since f is bounded, so f is Lebesgue integrable on $\mathcal{B}$ (see Theorem 9.13, [4]). Therefore, we get $|f|$ as Henstock-Kurzweil integrable. Thus $\theta(|f|)$ is Henstock-Kurzweil integrable.

Therefore $f \in \mathcal{H}^{-\theta}(\mu_\infty)$. So $HK(\mu_\infty) \subset \bigcup \mathcal{H}^{-\theta}(\mu_\infty)$ and hence the result. □

Corollary 6.1 $\mathcal{L}_1(\mu_\infty) = \bigcup \mathcal{H}^{-\theta}(\mu_\infty)$.

6.2 H-Orlicz Spaces

In this section we introduce the Henstock-Orlicz (shortly H-Orlicz) space over an arbitrary measure space $(\mathcal{B}, \Sigma, \mu_\infty)$ and discuss some topological properties of the space.

Definition 6.2 Let $(\mathcal{B}, \Sigma, \mu_\infty)$ be an arbitrary measure space and let $\mathcal{H}^{-\theta}(\mu_\infty)$ be the Henstock-Orlicz class on $(\mathcal{B}, \Sigma, \mu_\infty)$. Then the space $\mathcal{H}^{\theta}(\mu_\infty)$ of all measurable function $f : \mathcal{B} \to \mathbb{R}$ such that $\alpha f \in \mathcal{H}^{-\theta}(\mu_\infty)$ for some $\alpha > 0$ is called H-Orlicz space. We define the H-Orlicz space as follows. For some $\alpha > 0$

$$\mathcal{H}^{\theta}(\mu_\infty) = \left\{ f : \mathcal{B} \to \mathbb{R} \text{ measurable} : \int_{\mathcal{B}} \theta(\alpha f)d\mu_\infty \in \mathcal{H}(\mu_\infty)\right\}.$$

Proposition 6.2 $\mathcal{H}^{\theta}(\mu_\infty)$ *is a linear space.*

Proof Let $f_i \in \mathcal{H}^{\theta}(\mu_\infty)$, $i = 1, 2$. Then there exists α_i such that by definition $\alpha_i f_i \in \mathcal{H}^{-\theta}(\mu_\infty)$.

Let $\alpha = \min(\alpha_1, \alpha_2)$. Then for $\alpha > 0$.

$$\int_{\mathcal{B}} \theta\left(\frac{\alpha}{2}(f_1 + f_2)\right) d\mu_\infty \leq \frac{1}{2}\left[\int_{\mathcal{B}} \theta(\alpha_1 f_1) d\mu_\infty + \int_{\mathcal{B}} \theta(\alpha_2 f_2) d\mu_\infty\right].$$

Right-hand side is Henstock integrable. So, $f_1 + f_2 \in \mathcal{H}^\theta(\mu_\infty)$ as $\frac{\alpha}{2} > 0$.

If $f = f_1 = f_2$ implies $2f \in \mathcal{H}^\theta(\mu_\infty)$. So, $nf \in \mathcal{H}^\theta(\mu_\infty)$ for all integer $n > 1$. Therefore $\beta f \in \mathcal{H}^\theta(\mu_\infty)$ for any scalar β. Hence $\mathcal{H}^\theta(\mu_\infty)$ is a linear space. □

Definition 6.3 We define the Luxemburg norm on $\mathcal{H}^\theta(\mu_\infty)$ as follows:

$$H_\theta(f) = \inf\left\{a > 0 : HK \int_{\mathcal{B}} \theta\left(\frac{f}{a}\right) d\mu_\infty \leq 1\right\}.$$

It is understood that $\inf(\emptyset) = +\infty$. It is the gauge of the set

$$U = \left\{f \text{ measurable} : HK \int_{\mathcal{B}} \theta(f) d\mu_\infty \leq 1\right\}.$$

Proposition 6.3 *For each $f \in \mathcal{H}^\theta(\mu_\infty)$, there is an $\alpha > 0$ such that*

$$B_\theta = \left\{\alpha f = g \in \mathcal{H}^{-\theta}(\mu_\infty) : HK \int_{\mathcal{B}} \theta(g) d\mu_\infty \leq 1\right\}$$

is a solid subset of $\mathcal{H}^{-\theta}(\mu_\infty)$.

Proof Let $f, g \in \mathcal{H}^\theta(\mu_\infty)$. Then there exist $\alpha, \beta > 0$ such that $\alpha f, \beta g \in \mathcal{H}^{-\theta}(\mu_\infty)$. Let $\nu = \min(\alpha, \beta)$. Then for $\nu > 0$ and using the fact of convexity and monotonicity of θ, we get

$$\begin{aligned} &HK \int_{\mathcal{B}} \left(\frac{\nu}{2}(f + g)\right) d\mu_\infty \\ &\leq \frac{1}{2}\left[HK \int_{\mathcal{B}} \theta(\alpha f) d\mu_\infty) + HK \int_{\mathcal{B}} \theta(\beta g) d\mu_\infty\right]. \end{aligned}$$

The right side is Henstock-Kurzweil integrable. As $\frac{\nu}{2} > 0$, gives $f + g \in \mathcal{H}^\theta(\mu_\infty)$. In particular, with each f in $\mathcal{H}^\theta(\mu_\infty)$, $2f \in \mathcal{H}^\theta(\mu_\infty)$ and then $nf \in \mathcal{H}^\theta(\mu_\infty)$ for all integers $n > 1$, so that $\gamma f \in \mathcal{H}^\theta(\mu_\infty)$ for any scalar γ. Therefore the given set is solid and circled. To hold $\gamma f \in \mathcal{H}^\theta(\mu_\infty)$ for some $\gamma > 0$. Let $a_n \to 0$ be arbitrary and set $\gamma_n = \min(\gamma, a_n)$. Then $\gamma_n \to 0$ and $\theta(\gamma_n f) \leq \theta(\gamma f)$ and $\theta(\gamma_n f) \to 0$ as θ is a continuous Young function. Dominated convergence theorem gives us $(H) \int_{\mathcal{B}} \theta(\gamma_n f) \to 0$ so that for some n_0, we have $(H) \int_{\mathcal{B}} \theta(\gamma_{n_0} f) d\mu_\infty \leq 1$. Thus $\gamma_{n_0} f \in B_\theta$. □

Theorem 6.4 *The space $(\mathcal{H}^\theta(\mu_\infty), ||.||_{H_\theta})$ is normed linear space.*

Proof

(a) If $f = 0$ a.e., then $H_\theta(f) = 0$.
Conversely, if $H_\theta(f) = 0$.
Let $|f| > 0$, on a set of positive measure. If possible, then there exists a number $\delta > 0$ such that $A = \{p : |f(p)| \geq \delta\}$ satisfies $\mu_\infty(A) > 0$.
Now

$$\begin{aligned}\theta(n\delta)\mu_\infty(A) &= \int_A \theta(n\delta)d\mu_\infty \\ &\leq \int_A \theta(nf)d\mu_\infty \\ &\leq \int_{\mathcal{B}} \theta(nf)d\mu_\infty,\ n \geq 1.\end{aligned}$$

Since $\mu_\infty(A) > 0$ and $\theta(n\delta)d\mu_\infty \to \infty$ as $n \to \infty$. This is not possible so, $\mu_\infty(A) = 0$. Thus $f = 0$ a.e. Therefore $H_\theta(f) = 0$ if and only if $f = 0$ a.e.

(b) $H_\theta(\alpha f) = |\alpha| H_\theta(f)$, $\alpha \in \mathbb{R}$ or $\alpha \in \mathbb{C}$.
Let $\alpha \neq 0$.

$$\begin{aligned}H_\theta(\alpha f) &= \inf\left\{k > 0 : HK \int_{\mathcal{B}} \theta\left(\frac{\alpha f}{k}\right) d\mu_\infty \leq 1\right\} \\ &= |\alpha| \inf\left\{\frac{k}{|\alpha|} > 0 : HK \int_{\mathcal{B}} \theta\left(\frac{f}{\frac{k}{|\alpha|}}\right) d\mu_\infty \leq 1\right\} \\ &= |\alpha| H_\theta(f).\end{aligned}$$

(c) $H_\theta(f+g) \leq H_\theta(f) + H_\theta(g)$.
Let $f, g \in \mathcal{H}^\theta(\mu_\infty)$, $a > H_\theta(f)$ and $b > H_\theta(g)$. Then $0 < a < \infty$ and $0 < b < \infty$, and suppose $c = a + b > 0$. Since $f + g \in \mathcal{H}^\theta(\mu_\infty)$, so $H_\theta(f+g) < \infty$. Now

$$\begin{aligned}&HK \int_{\mathcal{B}} \theta\left(\frac{f+g}{c}\right) d\mu_\infty \\ &= HK \int_{\mathcal{B}} \theta\left[\frac{f}{a}.\frac{a}{c} + \frac{g}{b}.\frac{b}{c}\right] d\mu \\ &\leq \frac{a}{c} HK \int_{\mathcal{B}} \theta\left(\frac{f}{a}\right) d\mu_\infty + \frac{b}{c} HK \int_{\mathcal{B}} \theta\left(\frac{g}{b}\right) d\mu_\infty \\ &\leq \frac{a}{c} + \frac{b}{c} \leq 1.\end{aligned}$$

By Proposition 6.3, we have $\frac{1}{c}(f+g) \in B_\theta$ so that $H_\theta(f+g) \leq c = a+b$. If $a \to H_\theta(f)$ and $b \to H_\theta(g)$, we get the triangle inequality $H_\theta(f+g) \leq H_\theta(f) + H_\theta(g)$. Hence $(\mathcal{H}^\theta, ||.||_{H_\theta})$ is a normed linear space.

□

Lemma 6.1 *For $\mu_\infty(\mathcal{B}) < \infty$. Assume μ_∞ is bounded, then $\mathcal{H}^\theta(\mu_\infty) \subset \mathcal{L}_1(\mu_\infty)$, where the inclusion from $\mathcal{H}^\theta$ to $(\mathcal{L}_1, ||.||_1)$ is continuous.*

Proof We have for $u > 0$ and $v \geq 0$ such that $\theta(s) \geq us - v$ for all $s \geq 0$ implies $us \leq \theta(s) + v$.

Let $f \in \mathcal{H}^\theta(\mu_\infty)$. Then for small enough $a > 0$

$$\begin{aligned} a\int_{\mathcal{B}} |f| d\mu_\infty &\leq \frac{1}{u}\int_{\mathcal{B}} [\theta(af) + v] d\mu_\infty \\ &= \frac{1}{u}\int_{\mathcal{B}} \theta(af) d\mu_\infty + \frac{v\mu_\infty(\mathcal{B})}{u} \\ &< \infty. \end{aligned}$$

Therefore $f \in \mathcal{L}_1(\mu_\infty)$, so $\mathcal{H}^\theta(\mu_\infty) \subset \mathcal{L}_1(\mu_\infty)$ and this inclusion is continuous since with $c > 0$ taken sufficiently small for the inequality $\frac{u}{c} - v\mu_\infty(\mathcal{B}) \geq 1$ to hold, we have

$$\begin{aligned} HK\int_{\mathcal{B}} \theta\left(\frac{f}{c||f||_1}\right) d\mu_\infty &\geq \frac{u}{c} HK\int_{\mathcal{B}} \frac{f}{||f||_1} d\mu_\infty - v\mu_\infty(\mathcal{B}) \\ &= \frac{u}{c} - v\mu_\infty(\mathcal{B}) \geq 1. \end{aligned}$$

Hence $c||f||_1 \leq ||f||_{H_\theta}$. □

Corollary 6.2 $\mathcal{H}^\theta(\mu_\infty) \subset HK(\mu_\infty)$.

Lemma 6.2 *Let (f_n) be a sequence in $\mathcal{H}^\theta(\mu_\infty)$. Then the following are equivalent:*

(i) $\lim_{n\to\infty} ||f_n||_{H_\theta} = 0$.
(ii) For all $a > 0$, $\lim_{n\to\infty} \sup HK \int_{\mathcal{B}} \theta(af_n) d\mu_\infty \leq 1$.
(iii) For all $a > 0$, $\lim_{n\to\infty} HK \int_{\mathcal{B}} \theta(af_n) d\mu_\infty = 0$.

Proof For the proof of (i) implies (ii)

$$\begin{aligned} &\lim_{n\to\infty} ||f_n||_{H_\theta} = 0 \\ &\Longrightarrow \lim_{n\to\infty} \inf\left\{a > 0 : HK\int_{\mathcal{B}} \theta\left(\frac{f_n}{a}\right) d\mu_\infty \leq 1\right\} = 0 \\ &\Longrightarrow \inf\left\{a > 0 : \lim_{n\to\infty} HK\int_{\mathcal{B}} \theta\left(\frac{f_n}{a}\right) d\mu_\infty \leq 1\right\} = 0. \end{aligned}$$

Therefore, $\lim_{n\to\infty} \sup\ HK \int_{\mathcal{B}} \theta(af_n)d\mu_\infty \leq 1$.

For (iii) implies (ii) is obvious as θ is convex and $\theta(0) = 0$ for all $x \geq 0$ and $0 < \epsilon \leq 1$

$$\theta(x) = \theta(1-\epsilon)0 + \epsilon\left(\frac{x}{\epsilon}\right)$$
$$\leq (1-\epsilon)\theta(0) + \epsilon\theta\left(\frac{x}{\epsilon}\right).$$

Thus $\theta(x) \leq \epsilon\theta(\frac{x}{\epsilon}), x \geq 0,\ 0 < \epsilon \leq 1$.

From here (ii) implies (iii) follows easily.

For (ii) implies (i):

$$\lim_{n\to\infty} \sup HK \int_{\mathcal{B}} \theta(af_n)d\mu_\infty \leq 1$$
$$\implies \inf\left\{a > 0 : \lim_{n\to\infty} HK \int_{\mathcal{B}} \theta\left(\frac{f_n}{a}\right) d\mu_\infty \leq 1\right\} = 0$$
$$\implies \lim_{n\to\infty} \inf\left\{a > 0 : HK \int_{\mathcal{B}} \theta\left(\frac{f_n}{a}\right) d\mu_\infty \leq 1\right\} = 0$$
$$\implies \lim_{n\to\infty} ||f_n||_{H_\theta} = 0.$$

□

Theorem 6.5 *$(\mathcal{H}^\theta(\mu_\infty), ||.||_{H_\theta})$ is Banach space.*

Proof Let (f_n) be a Cauchy sequence in $\mathcal{H}^\theta(\mu_\infty)$. We find a countable partition $(E_k : k \geq 1\}$ of measurable subset of $\mathcal{B}$ such that $\mu_\infty(E_k) < \infty$ for all $k \geq 1$.

Now, in restriction $\mu_{\infty k}(.) = \mu_\infty(E_k \cap .)$.

As (f_n) is a Cauchy sequence in $\mathcal{H}^\theta(E_k, \mu_{\infty k})$. Using Lemma 6.1, (f_n) is a Cauchy sequence in $\mathcal{L}_1(\mu_\infty)$. As $\mathcal{L}_1(\mu_\infty)$ is complete, it is convergent in $\mathcal{L}_1(\mu_\infty)$ and one can extract a subsequence which converges μ_{∞_k}-a.e. pointwise to f in E_k. Using the diagonal extraction procedure, we can extract a subsequence (f_{n_k}) which converges μ_∞-a.e. pointwise to f on the whole space.

Let $a > 0$. Then there exist integer N_a and Lemma 6.2, $(H)\int_{\mathcal{B}} \theta(a(f_m - f_n))d\mu_\infty \leq 1$ for all $m, n \geq N_a$.

With Fatou's Lemma this gives for all $m \geq N_a$

$$HK \int_{\mathcal{B}} \theta(a(f_m - f))d\mu_\infty \leq \lim_{k\to\infty} \inf HK \int_{\mathcal{B}} \theta(a(f_m - f_{n_k}))d\mu_\infty \leq 1.$$

So, $f_m - f \in \mathcal{H}^\theta(\mu_\infty)$. But $f_m \in \mathcal{H}^\theta(\mu_\infty)$, so $f \in \mathcal{H}^\theta(\mu_\infty)$. Moreover,

$$\lim_{m\to\infty} \inf HK \int_{\mathcal{B}} \theta(a(f_m - f)d\mu_\infty \leq 1,\ \forall a > 0,$$

we have $\lim_{m\to\infty} ||f_m - f||_{\mathcal{H}_\theta} = 0$. So, it is a Banach space. □

Corollary 6.3 *$(\mathcal{H}^\theta(\mu_\infty), ||.||_{H_\theta})$ is a Banach space if $HK(\mu_\infty) = \mathcal{L}_1(\mu_\infty)$.*

From the Corollary 6.3, we can consider the Orlicz class as $I_\theta(f) = \int_0^\infty \theta(|f|)d\mu_\infty \in [0,\infty]$ and we set the Luxemburg norm as

$$||f||_{H_\theta} = \inf\left\{a > 0:\ I_\theta\left(\frac{f}{a}\right) \leq 1\right\} \in [0,\infty]$$

for every bounded measurable function $f : \mathbb{R}^+ \to \mathbb{R}$ with compact support. We put $\inf(\emptyset) = \infty$. We will find the following proposition directly from the convexity of θ.

Proposition 6.4 *Let f, g be bounded measurable functions and $0 < \alpha < 1$. Then*

1. $I_\theta(\alpha f + (1-\alpha)g) \leq \alpha I_\theta(f) + (1-\alpha)I_\theta(g)$.
2. $I_\theta(\alpha f) \leq \alpha I_\theta(f)$.
3. *The set $B_\theta = \{f \in \mathcal{H}_\theta(\mu_\infty):\ I_\theta(f) \leq 1\}$ is convex.*
4. *If $f \in I_\theta$ and $||f||_{H_\theta} \leq 1$, then $f \in B_\theta$.*

Proof Proofs are similar with the proof of Proposition 13.2.1 [8]. □

Theorem 6.6 *The space $(\mathcal{H}^\theta(\mu_\infty), ||.||_{H_\theta})$ is a symmetric space.*

Proof If f is bounded measurable function, then

$$\begin{aligned}
H_\theta(f) &= HK\int_{\mathcal{B}} \theta(|f|)d\mu_\infty \\
&= HK\int_{\mathcal{B}} \eta_{\theta o|f|}d\mu_\infty \\
&= HK\int_{\mathcal{B}} \mu_\infty\{\theta(|f|) > x\}d\mu_\infty \\
&= HK\int_{\mathcal{B}} \mu_\infty\{|f| > \theta^{-1}(x)\}d\mu_\infty \\
&= HK\int_{\mathcal{B}} \mu_\infty\{|f| > y\}\theta(y) \\
&= HK\int_{\mathcal{B}} \eta_{|f|}d\theta.
\end{aligned}$$

Let $f \in \mathcal{H}^{\theta}(\mu_\infty)$ and let g be equimeasurable to f, that is $\eta_{|f|} = \eta_{|g|}$. Then we have

$$\begin{aligned} H_\theta(f) &= HK \int_{\mathcal{B}} \eta_{|f|} d\theta \\ &= HK \int_{\mathcal{B}} \eta_{|g|} d\theta \\ &= H_\theta(g). \end{aligned}$$

Therefore, $g \in \mathcal{H}^{\theta}(\mu_\infty)$, so $||f||_{H_\theta} = ||g||_{H_\theta}$. Hence, this space is symmetric.

It is easy to check that $(\mathcal{H}^{\theta}(\mu_\infty), ||.||_{H_\theta})$ is a normed ideal lattice. □

Lemma 6.3 *For any bounded measurable function g on $\mathcal{B}$, $||g||_{H_\theta} = 0$ if and only if $g = 0$ a.e.*

Proof As $||g||_{H_\theta} = 0$ if and only if $\inf\{a > 0 : (H)\int_{\mathcal{B}} \theta(\frac{g}{a}) d\mu_\infty \leq 1\} = 0$. This gives

$$\begin{aligned} ||g||_{H_\theta} = 0 &\iff HK \int_{\mathcal{B}} \theta\left(\frac{g}{a}\right) d\mu_\infty = 0 \text{ for all } a > 0 \text{ (by Lemma 6.2(iii))} \\ &\iff \theta\left(\frac{g}{a}\right) = 0 \; a.e. \text{ for all } a > 0 \\ &\iff g = 0 \; \mu_\infty - a.e. \end{aligned}$$

□

Lemma 6.4 *If $0 < ||f||_{H_\theta} < \infty$, then $(H)\int_{I \subseteq \mathcal{B}} \theta\left(\frac{f}{||f||_{H_\theta}}\right) \leq A$, where $A = \lim S\left(\theta\left(\frac{f}{||f||_{H_\theta}}\right), D_k\right)$ and D_k is a sequence of tagged partition.*

Proof Let $||f||_{H_\theta} > 0$. Then $\int_{\mathcal{B}} \theta\left(\frac{f}{||f||_{H_\theta}}\right) d\mu_\infty$ is Henstock-Kurzweil integrable. By (Ch 1, Theorem 6 [9]), for every k there exists a gauge δ_k on $I \subseteq \mathcal{B}$ such that $D_1, D_2 \leq \delta_k$, then

$$\left| S\left(\theta\left(\frac{f}{||f||_{H_\theta}}\right), D_1\right) - S\left(\theta\left(\frac{f}{||f||_{H_\theta}}\right), D_2\right) \right| < \frac{1}{k}.$$

For each k, let $D_k < \delta_k$. If $k > j$, we have

$$\left| S\left(\theta\left(\frac{f}{||f||_{H_\theta}}\right), D_k\right) - S\left(\theta\left(\frac{f}{||f||_{H_\theta}}\right), D_j\right) \right| < \frac{1}{j}.$$

Thus $\left\{ S\left(\theta\left(\frac{f}{||f||_{H_\theta}}\right), D_k\right) \right\}$ is a Cauchy sequence in $\mathbb{R}$.

Let $A = \lim_k S\left(\theta\left(\frac{f}{||f||_{H_\theta}}\right), D_k\right)$. Then
$\left|S\left(\theta\left(\frac{f}{||f||_{H_\theta}}\right), D_k\right) - A\right| \leq \frac{1}{k}$ for all k.
Let $\epsilon > 0$. There exists natural number $'n'$ such that $\frac{1}{n} < \frac{\epsilon}{2}$.
Assume $D < \delta_n$. Then

$$\begin{aligned}
&\left|S\left(\theta\left(\frac{f}{||f||_{H_\theta}}\right), D\right) - A\right| \\
&\leq \left|S\left(\theta\left(\frac{f}{||f||_{H_\theta}}\right), D\right) - S\left(\theta\left(\frac{f}{||f||_{H_\theta}}\right), D_n\right)\right| \\
&+ \left|S\left(\theta\left(\frac{f}{||f||_{H_\theta}}\right), D_n\right) - A\right| \\
&< \frac{1}{n} + \frac{1}{n} < \epsilon.
\end{aligned}$$

Hence $HK \int_{I \subseteq \mathcal{B}} \theta\left(\frac{f}{||f||_{H_\theta}}\right) d\mu \leq A$. □

Theorem 6.7 *For all $f \in \mathcal{H}^\theta(\mu_\infty)$, $g \in \mathcal{H}^\Phi(\mu_\infty)$, where Φ is complementary function of θ, then*

$$HK \int_{\mathcal{B}} |fg| d\mu_\infty \leq ||f||_{H_\theta} ||g||_{H_\Phi}.$$

Proof If $||f||_{H_\theta} = 0$, $||g||_{H_\Phi} = 0$. Then by Lemma 6.3, gives the completeness of the result. Let $0 < ||f||_{H_\theta}$, $0 < ||g||_{H_\Phi}$. By Young's inequality, we get

$$st \leq \theta(s) + \Phi(t); \ s, t \geq 0.$$

Let $s = \frac{f}{||f||_{H_\theta}}$, $t = \frac{g}{||g||_{H_\Phi}}$. Then we have

$$\begin{aligned}
&HK \int_{\mathcal{B}} \frac{f}{||f||_{H_\theta}} \frac{g}{||g||_{H_\Phi}} d\mu_\infty \\
&\leq HK \int_{\mathcal{B}} \theta\left(\frac{f}{||f||_{H_\theta}}\right) d\mu_\infty + HK \int_{\mathcal{B}} \Phi\left(\frac{g}{||g||_{H_\Phi}}\right) d\mu_\infty \\
&\leq A + A.
\end{aligned}$$

Therefore $\int_{\mathcal{B}} fg d\mu_\infty \leq 2A||f||_{H_\theta}||g||_{H_\Phi}$.
In particular $A = \frac{1}{2}$ gives $HK \int_\Omega |fg| d\mu_\infty \leq ||f||_{H_\theta}||g||_{H_\Phi}$. □

In the H-Orlicz space, functions are bounded measurable with compact support. This is only the fact that Henstock integral is Lebesgue integral. Though $\mathcal{L}_1(\mu_\infty) \subseteq HK(\mu_\infty)$ in this assumption $\mathcal{L}_1(\mu_\infty) = HK(\mu_\infty)$.

Theorem 6.8 *Let θ be a continuous Young function with $\theta(x) = 0$ if and only if $x = 0$, with $\theta \in \Delta_2$. Then the space $\mathcal{H}^\theta(\mu_\infty)$ is separable if and only if the measure μ_∞ is separable.*

Proof The proof follows from the similar technique used in (Proposition 2, [7], p 91). □

6.3 Denseness of $C_0^\infty(\mathbb{R}^n)$

Lemma 6.5 *Let $\mathbb{M}$ be an element of $C_0^\infty(\mathbb{R}^n)$ satisfying $\mathbb{M} \geq 0$ and $\int \mathbb{M}(t) d\mu_\infty(t) = 1$. Define a sequence $(\mathbb{M}_n)_{n\geq 1}$ of elements of $C_0^\infty(\mathbb{R}^n)$ by $\mathbb{M}_n(t) = n\mathbb{M}(nt)$. Let θ be an N-function and $f \in \mathcal{H}^\theta(\mu_\infty)$, then the convolution $\mathbb{M}_n * f \in \mathcal{H}^\theta(\mu_\infty)$ and $||\mathbb{M}_n * f - f||_{H_\theta} \to 0$ as $n \to \infty$.*

Proof Let Φ be the complementary N-function of θ and let $g \in \mathcal{H}^\Phi(\mu_\infty)$ with $||g||_{H_\Phi} = 1$.

Then

$$
\begin{aligned}
& HK \int |\mathbb{M}_n * f(x) - f(x)||g(x)| d\mu_\infty(x) \\
&= HK \int |\mathbb{M}_n * f(x) - HK \int \mathbb{M}(t) d\mu_\infty(t) f(x)||g(x)| d\mu_\infty(x) \\
&= HK \int | \int \mathbb{M}_n(t) f(x-t) d\mu_\infty(t) \\
&\quad - HK \int \mathbb{M}(t) d\mu(t) f(x)||g(x)| d\mu_\infty(x) \\
&\leq HK \int | \int \mathbb{M}_n(t) f(x-t) d\mu_\infty(t) \\
&\quad - HK \int \mathbb{M}_n(t) d\mu_\infty(t) f(x)||g(x)| d\mu_\infty(x) \\
&\leq HK \int \left\{ \int |f(x-t) - f(x)||g(x)| d\mu_\infty(x) \right\} \mathbb{M}_n(t) d\mu_\infty(t) \\
&\leq 2HK \int ||f_t - f||_{H_\theta} \mathbb{M}_n(t) d\mu_\infty(t) \text{ (by Theorem 6.7)},
\end{aligned}
$$

where $f_t(x) = f(x-t)$.

By assumption $||g||_{H_\Phi} = 1$, therefore by Theorem 6.7, we have

$$\begin{aligned} ||\mathbb{M}_n * f - f||_{H_\theta} &\leq 2HK \int ||f_t - f||_{H_\theta} \mathbb{M}_n(t) d\mu_\infty(t) \\ &= 2HK \int ||f_{\frac{t}{n}} - f||_{H_\theta} \mathbb{M}(t) d\mu_\infty(t). \end{aligned}$$

Since $f \in \mathcal{H}^\theta(\mu_\infty)$ and $\mathbb{M}$ has compact support, so for every $\epsilon > 0$ there exists k sufficiently large such that

$$\begin{aligned} HK \int ||f_{\frac{t}{n}} - f||_{H_\theta} \mathbb{M}(t) d\mu_\infty(t) &\leq \epsilon HK \int \mathbb{M}(t) d\mu_\infty(t) \\ &= \epsilon. \end{aligned}$$

Therefore $||\mathbb{M}_n * f - f||_{H_\theta} \to 0$ as $n \to \infty$. It is clear that $\mathbb{M}_n * f \in \mathcal{H}^\theta(\mu_\infty)$. □

Theorem 6.9 *If $\mathcal{B} = \mathbb{R}^n$, then $C_0^\infty(\mathbb{R}^n)$ is dense in $\mathcal{H}^\theta(\mathbb{R}^n) = \mathcal{H}^\theta(\mathbb{R}^n, \mu_\infty)$.*

Proof Suppose $f \in \mathcal{H}^\theta(\mu_\infty)$ has compact support. Let us assume a sequence (M_n) of elements of C_0^∞ such that $M_n(x) = 1$ for $x \leq n$ and $M_n(x) = 0$ for $x \geq 2n$.

Clearly $M_n * f$ is a function with compact support, and $M_n * f \to f \in \mathcal{H}^\theta(\mathbb{R}^n)$. Let (M_n) be the sequence defined in Lemma 6.5. Define

$$\begin{aligned} f_n(x) &= f * M_n(x) \\ &= HK \int f(t) M_n(x - t) d\mu_\infty(t). \end{aligned}$$

Derivative of distributional function is again distributional function, (page 33 [1] and Lemma 5.7 of [3]) shows that $D^\alpha f \in \mathcal{H}^\theta(\mathbb{R}^n)$. So, for all α, $|\alpha| \leq m$ and $M_n \in C_0^\infty$.

We have

$$\begin{aligned} D^\alpha f_n(x) &= D^\alpha HK \int f(t) M_n(x - t) d\mu_\infty(t) \\ &= HK \int f(t) D^\alpha M_n(x - t) d\mu_\infty(t) \\ &= HK \int D^\alpha f(t) M_n(x - t) d\mu_\infty(t). \end{aligned}$$

So, $||D^\alpha f_n - D^\alpha f||_{H_\theta} = ||D^\alpha f * M_n - D^\alpha f||_{H_\theta} \to 0$ as $n \to \infty$. Therefore, $C_0^\infty(\mathbb{R}^n)$ is dense in $\mathcal{H}^\theta(\mathbb{R}^n, \mu_\infty)$. □

References

1. A-Algwaiz, M.: Theory of Distributions. Pure and Applied Mathematics, Monograph. Marcel Dekker (1992)
2. Alexiewicz, A.: Linear functionals on Denjoy-integrable functions. Collectanea Mathematica **1**, 289–293 (1948)
3. Duistermaat, J.J., Kolk, J.A.C.: Distribution: Theory and Application. Springer, Berlin Heidelberg, New York (2006)
4. Gordon, R.: The Integrals of Lebesgue, Denjoy, Perron, and Henstock. American Mathematical Society, Providence (1994)
5. Krasonsel'skii, M.A., Rutickii, Ya.B.: Convex Functions and Orlicz Spaces. P. Noordfoff, Gronin-gen (1961)
6. Lindenstrauss, J., Tzafriri, L.: Classical Banach Spaces II, Function Spaces. A Series of Modern Surveys in Mathematics. Springer, Berlin-Heidelberg-New York (1979)
7. Rao, M.M., Ren, Z.: The Theory of Orlicz Spaces. Marcel Dekker, New Work (1991)
8. Rubshtein, B.Z.A., Grabarnik, G.Y., Muratov, M.A., Pashkova, Y.S.: Foundations of Symmetric Spaces of Measurable Functions. Developments in Mathematics. Springer, Berlin (2016)
9. Swartz, C.: Introduction to Gauge Integral. World Scientific, Singapore (2001)

Chapter 7
Geometrical Properties of Henstock-Orlicz Spaces

7.1 Introduction

Throughout this chapter, Σ denotes a σ-algebra of subsets of an abstract set $T \neq \emptyset$. $\mathcal{P}(T)$ be the class of all subset of T, $\Sigma \subset \mathcal{P}(T)$ be a σ-algebra, $\mathcal{X}$ be a Banach space, and $\mathcal{X}'$ be its topological dual. For each $A \in \Sigma$, the characteristic function of A

$$\chi_A(t) = \begin{cases} 1 \text{ if } t \in A, \\ 0 \text{ if } t \in T \setminus A. \end{cases}$$

Recalling a vector measure is a σ-additive set function $\mu_\infty : \Sigma \to \mathcal{X}$, the σ-additivity of μ is equivalent to the σ-additivity of the scalar-valued set functions $x'\mu_\infty : A \to x'(\mu_\infty(A))$ on Σ for every $x' \in \mathcal{X}'$.

Definition 7.1 The variation $|\mu_\infty|$ of μ_∞ is defined by

$$|\mu_\infty|(A) = \sup\left\{\sum_{i=1}^{r} ||\mu_\infty(A_i)|| \ : A_i \in \Sigma,\ i = 1, 2, .., r; A_i \cap A_j = \emptyset \right.$$
$$\left. \text{for } i \neq j;\ \bigcup_{i=1}^{r} A_i \subset A \right\}.$$

The semivariation $||\mu_\infty||$ of μ_∞ by

$$||\mu_\infty||(A) = \sup_{x' \in \mathcal{X}', ||x'|| \leq 1} |x'\mu_\infty|(A).$$

B. Hazarika, H. Kalita, *Theory of Henstock–Orlicz Spaces*, Mathematical Marvels: Texts and Monographs in the Spirit of CR Rao,
https://doi.org/10.1007/978-981-96-9548-5_7

A function $f : T \to \mathbb{R}$ is said to be μ_∞-measurable if

$$f^{-1}(B) \cap \{t \in T : f(t) \neq 0\} \in \Sigma$$

for each Borel subset $\mathfrak{B} \subset \mathbb{R}$.

Definition 7.2 We say that a Σ-measurable function $f : T \to \mathbb{R}$ is Kluvánek-Lewis-Henstock-Kurzweil μ_∞-integrable , shortly (HKL) μ_∞-integrable if the following properties hold:

$$f \text{ is } |x'\mu_\infty| - \text{Henstock-Kurzweil integrable for each } x' \in \mathcal{X}',$$

and for every $A \in \Sigma$ there is $x_A^{(H)} \in \mathcal{X}$ with

$$x'(x_A^{(H)}) = HK \int_A f \, d|x'\mu_\infty| \text{ for all } x' \in \mathcal{X}', \tag{7.1}$$

where the symbol "HK'' denotes the usual Henstock-Kurzweil integral of a real-valued function with respect to an (extended) real-valued measure.

Definition 7.3 [9] A function $f : T \to \mathcal{X}$ is M-measurable if a sequence of simple functions from $\mathcal{S}_{\mathcal{X}}(\Sigma)$ converges to f M-a.e.

M denotes a σ-bounded family of positive measures defined on Σ. This means for each $E \in \Sigma$, there exists a pairwise disjoint collection $\{E_i\}_{i=1}^{\infty}$, $E_i \in \Sigma$ such that $E = \bigcup_{i=1}^{\infty} E_i$ and $\mathcal{S}_{\mathcal{X}}(E)$ is a set of all $\mathcal{X}$ valued simple functions. If f is M-measurable, then f is μ_∞-measurable for each $\mu_\infty \in M$, we define $M(f) = \sup_{\mu_\infty \in M} HK \int_T ||f|| d|x'\mu_\infty|$. In this chapter we consider all functions are M-measurable. Let $\mathcal{M}_{\mathcal{X}}(M)$ denote the space of all M-measurable functions $f : T \to \mathcal{X}$ for which $M(f) < \infty$, then $\mathcal{M}_{\mathcal{X}}(M)$ is complete with respect to the semi norm $M(.)$. If $M(f) < \infty$, then f is integrable in $\mathcal{M}_{\mathcal{X}}(E)$.

7.2 Henstock-Orlicz Spaces and Vector Measure

If $f \in \mathcal{H}^\theta(|x'\mu_\infty|)$, then there is a $k > 0$ such that $\theta(\frac{f}{k}) \in HK(|x'\mu_\infty|)$, where $HK(|x'\mu_\infty|)$ is Henstock-Kurzweil integrable space. The space $\mathcal{H}^\theta(|x'\mu_\infty|)$ endowed with the Luxemburg norm

$$||f||_{\mathcal{H}^\theta} = \inf\left\{\alpha > 0 : HK \int_T \theta\left(\frac{f}{\alpha}\right) d|x'\mu_\infty| \leq 1\right\} \tag{7.2}$$

is a Banach space.

For complementary Young function θ and ϕ we define

$$\mu_{\infty_\phi}(g) = HK \int_T \phi(|g|) d\mu_\infty, \ g \in \mathcal{S}(\Sigma)$$

and $M_\phi(g) = \sup\limits_{\mu_\infty \in \mathcal{M}} \mu_{\infty_\phi}(g)$.

Definition 7.4 For an M-measurable function $f : T \to \mathcal{X}$, let us define

$$||f||_{\theta,M} = \sup_{\mu_\infty \in M} \sup_{g \in \mathcal{S}(\Sigma)} \left\{ HK \int_T ||fg|| d|x'\mu_\infty| \ : M_\phi(g) \leq 1 \right\} \quad . \quad (7.3)$$

Clearly (7.2) and (7.3) are equivalent.

Definition 7.5 [10] If $\mathcal{X}$ and Y are two Banach spaces and $\mu_\infty : \Sigma \to L(\mathcal{X}, Y)$ a countable additive measure, then (θ, M)-variation of μ_∞ is defined by

$$|\mu_\infty|_{\theta,M}(E) = \sup \sum_{i=1}^{n} ||\mu_\infty(E_i) x_i|| \text{ for each } E \in \Sigma.$$

Here the supremum is taken over all simple functions $f = \sum\limits_{i=1}^{n} x_i ch(E_i)$ belonging to $\mathcal{S}_\mathcal{X}(\Sigma)$ with pairwise disjoint $E_i \subset E$ and $M_\theta(f) \leq 1$. The (θ, M) semivariation of μ_∞, for $E \in \Sigma$ by

$$||\mu_\infty||_{\theta,M}(E) = \sup || \sum_{i=1}^{n} \mu_\infty(E_i) x_i ||,$$

where the supremum is taken over all simple functions $f = \sum\limits_{i=1}^{n} x_i \chi_{E_i}$ belonging to $\mathcal{S}_\mathcal{X}(\Sigma)$ with pairwise disjoint $E_i \subset E$ and $M_\theta(f) \leq 1$.

Theorem 7.1 *$\mathcal{H}^\theta(M, \mathcal{X})$ is complete with respect to $||.||_{\theta,M}$.*

Proof Suppose $(f_n)_{n=1}^\infty$ is a Cauchy sequence in $\left(\mathcal{H}^\theta(M, \mathcal{X}), ||.||_{\theta,M}\right)$, then $||f_n - f_m||_{\theta,M} \leq \frac{1}{2}$ for $m > n$. In general n_k, $k \in \mathbb{N}$ such that $||f_n - f_m||_{\theta,M} < \frac{1}{2^k}$ for $m > n$. Since M is a σ-bounded, we can write $T = \bigcup\limits_{i=1}^{\infty} E_i$, $E_i \cap E_j = \emptyset$ for $i \neq j$ and $M(E_i) < \infty$ for each $i = 1, 2, \ldots$, let $t > 0$ be such that $M(E_i)\phi(t) \leq 1$. If $g(x) = t$ on E_i and $g(x) = 0$ otherwise, then we have $M_\phi(g) \leq 1$ and consequently $M(f_n - f_m) < \epsilon$ for $n, m \geq M_0$, so $M(f_n - f_m) < \frac{\epsilon}{t}$ for $n, m \geq M_0$. This gives $\{f_n\}$ is a Cauchy sequence in $\mathcal{S}_\mathcal{X}(M)$. As (f_n) is Cauchy sequence in $\mathcal{S}_\mathcal{X}(M)$, then there exists a positive integer M_1 and a function $f_{M_1} \in \mathcal{S}_\mathcal{X}(M)$ such that $M(f_n - f_{M_1}) < \frac{1}{2}$ for $n \geq M_1$. Similarly on E_2, $M(f_n - f_{M_2}) < \frac{1}{2^2}$ for $n \geq M_2$

and so on. The series $M(f_{M_1}) + M(f_{M_2} - f_{M_1}) + \dots$ is convergent. The fact $\mathcal{S}_X(M)$ is complete gives $(f_{M_1}) + (f_{M_2} - f_{M_1}) + \dots$ converges a.e. to a function, say f in $\mathcal{S}_X(M)$, so, $M(f_n - f)(g) < \epsilon$ whenever $n \geq M_0$ and $M_\phi(g) \leq 1$. Thus $f_n - f \in \mathcal{H}^\theta(M, X)$ we have $f \in \mathcal{H}^\theta(M, X)$. So $\left(\mathcal{H}^\theta(M, X), ||.||_{\theta,M}\right)$ is complete. □

Theorem 7.2 *Let* $f_n \in \mathcal{H}^\theta(M, X)$. *For* $n = 1, 2, \dots$ *then the following are equivalent:*

1. $f \in \mathcal{H}^\theta(M, X)$ *and* $||f_n - f||_{\theta,M} \to 0$ *as* $n \to \infty$.
2. $||f_n - f_m||_{\theta,M} \to 0$ *as* $m, n \to \infty$ *and hence there exists a subsequence* (f_{n_k}) *of* (f_n) *such that* $f_{n_k} \to f$ M*-a.e.*

Proof Proof is similar as [9, Theorem 3.3]. □

Definition 7.6 We denote $HK(M, X)$ is the collection of all functions $f : T \to X$ which are M-measurable and for which $\theta(||f||) \in HK(M, \mathbb{R})$, where $HK(M, \mathbb{R})$ is the Henstock-Kurzweil integrable function space.

Theorem 7.3 *If* $f \in \mathcal{H}^\theta(M, X)$, *there exists a constant* $N > 0$ *such that* $Nf \in HK(M, X)$. *Moreover if* $f \in \mathcal{H}^\theta(M, X)$ *and* $f(x) \neq 0$, *then*

$$\sup_{\mu_\infty \in M} HK \int_T \theta\left(\frac{||f||}{||f||_{\theta,M}}\right) d|x'\mu_\infty| \leq 1.$$

Proof Let $f \in \mathcal{H}^\theta(M, X)$ and $g \in \mathcal{S}(\Sigma)$, we need to prove for each $\mu_\infty \in M$,

$$HK \int_T \frac{||fg||}{||f||_{\theta,M}} d|x'\mu_\infty| \leq \max\{1, \mu_{\infty_\phi}(g)\}.$$

If $\alpha > 1$ and $t > 0$ then $\theta(\alpha t) \geq \alpha\phi(t)$. So, let $f \in \mathcal{H}^\theta(M, X)$ and $g \in \mathcal{S}(\Sigma)$ then

$$HK \int_T ||fg|| d|x'\mu_\infty| \leq ||f||_{\theta,M} \text{ provided } \mu_{\infty_\phi} \leq 1. \tag{7.4}$$

If $1 < \mu_{\infty_\phi}(g) < \infty$, then

$$\begin{aligned}\phi(|g|) &= \phi\left(\mu_{\infty_\phi} \frac{|g|}{\mu_{\infty_\phi}(g)}\right) \\ &\geq \mu_{\infty_\phi}(g)\phi\left(\frac{|g|}{\mu_{\infty_\phi}(g)}\right).\end{aligned}$$

That is $\frac{\phi(|g|)}{\mu_{\infty_\phi}(g)} \geq \phi\left(\frac{|g|}{\mu_{\infty_\phi}(g)}\right)$. So,

$$HK\int_T \phi\left(\frac{|g|}{\mu_{\infty_\phi}(g)}\right) d|x'\mu_\infty| \leq HK\int_T \frac{\phi(|g|)}{\mu_{\infty_\phi}(g)} d|x'\mu_\infty|$$
$$\leq 1.$$

So, $HK\int_T \frac{||fg||}{\mu_{\infty_\phi}(g)} d|x'\mu_\infty| \leq ||f||_{\theta,M}$. This means

$$HK\int_T ||fg|| d|x'\mu_\infty| \leq ||f||_{\theta,M}\mu_{\infty_\phi}(g). \tag{7.5}$$

From (7.4) and (7.5) we get

$$HK\int_T ||fg|| d|x'\mu_\infty| \leq ||f||_{\theta,M} \max\{1, \mu_{\infty_{\phi}(g)}\}$$
$$i.e., HK\int_T \frac{||fg||}{||f||_{\theta,M}} d|x'\mu_\infty| \leq \max\{1, \mu_{\infty_\phi}(g)\}.$$

Next suppose that $f \in \mathcal{H}^\theta(M, \mathcal{X})$ and $||f||_{\theta,M} > 0$, from here we get two cases.

Case 1: If f is bounded and has support E (say), of finite M-measurable sets. Let $g = \frac{f}{||f||_{\theta,M}}$. Since g is bounded on E, $\phi(\chi(||g||))$ is bounded, so $\sup_{\mu_\infty \in M} HK\int_T \phi(\chi(||g||)) d|x'\mu_\infty|$ exists. Using Young's inequality

$$\sup_{\mu_\infty \in M} HK\int_T \theta(||g||) d|x'\mu_\infty|$$
$$= \sup_{\mu_\infty \in M}\Big\{HK\int_T \theta(||g||) d|x'\mu_\infty| + HK\int_T \phi(\chi(||g||)) d|x'\mu_\infty|$$
$$-HK\int_T \phi(\chi(||g||)) d|x'\mu_\infty|\Big\}$$
$$= \sup_{\mu_\infty \in M}\left\{HK\int_T ||g\chi(||g||) d|x'\mu_\infty| - HK\int_T \phi(\chi(||g||)) d|x'\mu_\infty|\right\}.$$

As $\chi(||g||)$ is positive bounded M-measurable function, there exists a nondecreasing sequence of positive simple functions $\{g_n\}$ such that $g_n \to \chi(||g||)$ M-a.e. and $g_n \leq \chi(||g||)$ and $||gg_n|| \leq ||g\chi(||g||)||$. So,

$$\sup_{\mu_\infty \in M} HK\int_T ||g\chi(||g||)|| d|x'\mu_\infty| \leq \sup_{\mu_\infty \in M} \lim_n HK\int_T ||gg_n|| d|x'\mu_\infty|.$$

Hence

$$\sup_{\mu_\infty \in M} HK \int_T \theta(||g||) d|x'\mu_\infty| \leq \sup_{\mu_\infty \in M} \max\{1 - \mu_{\infty_\phi}(\chi ||g||, 0)$$
$$\leq 1.$$

Case 2: Suppose $f \in \mathcal{H}^\theta(M, X)$ and $||f||_{\theta,M} > 0$. Since M is σ-bounded, we can choose an increasing sequence of sets $\{E_n\}$ of finite M-measurable such that $T = \bigcup_{n=1}^{\infty} E_n$. Using (7.2) for $M(||f_n|| \to M||f||)$ for each $n = 1, 2, \ldots,$ we have

$$\sup_{\mu_\infty \in M} HK \int_T \theta\left(\frac{||f_n||}{||f_n||_{\theta,M}}\right) d|x'\mu_\infty| \leq 1.$$

□

Proposition 7.1 *Suppose $\theta \in \Delta_2$, then $\mathcal{H}^\theta(M, \mathcal{X}) = H(M, \mathcal{X})$ also if $f_n \in \mathcal{H}^\theta(M, \mathcal{X})$ with $\sup_{\mu_\infty \in M} (H) \int_T \theta(||f_n||) d|x'\mu_\infty| \to 0$, then $||f_n||_{\theta,M} \to 0$.*

Proof Let $f : T \to X$ be M-measurable function and $g \in \mathcal{S}(\Sigma)$, then by Young's inequality

$$HK \int_T ||fg|| d|x'\mu_\infty| \leq (H) \int_T \theta\left(||f|| d|x'\mu_\infty|\right).$$

So, if $f \in HK(M, \mathcal{X})$, then $f \in \mathcal{H}^\theta(M, \mathcal{X})$. Let $f \in \mathcal{H}^\theta(M, \mathcal{X})$ and $||f||_{\theta,M} \neq 0$. Now from Theorem 7.3, $\frac{f}{||f||_{\theta,M}} \in HK(M, \mathcal{X})$. Now as $f \in \mathcal{H}^\theta(M, \mathcal{X})$ implies $||f||_{\theta,M} < \infty$ that is $||f||_{\theta,M} \leq 2^m$; $m > 0$. Therefore

$$\theta(||f||) = \theta\left(\frac{||f||}{||f||_{\theta,M}}\right) ||f||_{\theta,M}$$
$$\leq 2^m \theta\left(\frac{||f||}{||f||_{\theta,M}}\right).$$

Thus, $\theta(||f||) \in HK(M, \mathcal{X})$, so $f \in HK(M, \mathcal{X})$). Hence $\mathcal{H}^\theta(M, \mathcal{X}) = HK(M, \mathcal{X})$. □

Theorem 7.4 *$\mathcal{S}_X(\Sigma)$ is dense in $\mathcal{H}^\theta(M, \mathcal{X})$.*

Proof Choose arbitrarily $\epsilon > 0$ and $f \in \mathcal{H}^\theta(M, \mathcal{X})$ that is $g \in HK(M, \mathcal{X})$ with

$$||g - f||_{\mathcal{H}^\theta(M,\mathcal{X})} \leq \frac{\epsilon}{N+1}.$$

Moreover, in correspondence with ϵ and g, we find a Σ-simple function s, with

$$||s - g||_{H(M,\mathcal{X})} \leq \frac{\epsilon}{N+1}$$

As $||.||_{\mathcal{H}^\theta(M,\mathcal{X})} \leq M||.||_{HK(M,\mathcal{X})}$, and hence, we obtain

$$\begin{aligned}
||s - f||_{\mathcal{H}^\theta(M,\mathcal{X})} &\leq ||s - f||_{\mathcal{H}^\theta(M,\mathcal{X})} + ||g - f||_{\mathcal{H}^\theta(M,\mathcal{X})} \\
&\leq N(||s - g||_{HK(M,X)} + ||g - f||_{\mathcal{H}^\theta(M,\mathcal{X})}) \\
&\leq \frac{N\epsilon}{N+1} + \frac{\epsilon}{N+1} \\
&= \epsilon.
\end{aligned}$$

So, $\mathcal{S}_\mathcal{X}(\Sigma)$ is dense in $\mathcal{H}^\theta(M, \mathcal{X})$. □

Proposition 7.2 *Suppose* $\mu_\infty : \Sigma \to L(\mathcal{X}, Y)$ *is countable additive with* $||\mu_{\infty_{\mathcal{X},Y}}||(T) < \infty$, *then* $\mathcal{H}^\theta(M, \mathcal{X}) \subset HK(\mu_{\infty_{\mathcal{X},Y}}, \mathcal{X})$ *and* $\mu_{\infty_{\mathcal{X},Y}}(f) \leq ||\mu_{\infty_{\theta,M}}||(T)||f||_{\theta,M}$ *for* $f \in \mathcal{H}^\theta(M, \mathcal{X})$.

Proof Let $f \in \mathcal{H}^\theta(M, \mathcal{X})$. Then by Theorem 7.4, there is a sequence $\{f_n\}$ of simple functions converging to f in $\mathcal{H}^\theta(M, \mathcal{X})$. Let $y_1 \in Y'$ be such that $|y_1| \leq 1$. If $f = \sum_{i=1}^{n} x_i ch(E_i) \in \mathcal{S}_\mathcal{X}(E)$, then

$$HK \int_T ||f||d|\mu_{\infty_i} = \sum_{i=1}^{n} ||x_i|||\mu_{\infty_{y_1}}|(E_i).$$

Suppose $\epsilon > 0$ be arbitrary and for each i, let $\{\mathcal{F}_{ij}\}$ be a finite family of disjoint sets in Σ contained in E_i such that

$$|\mu_{\infty_{y_i}}|(E_i) < \sum_{ij} ||\mu_{\infty_{y_1}}(\mathcal{F}_{ij})|| + \frac{\epsilon}{n||x_i||}.$$

Then

$$HK \int_T ||f||d|\mu_{\infty_{y_1}}| < \sum_{ij} ||\mu_{\infty_{y_1}}(\mathcal{F}_{ij})||||x_i|| + \epsilon.$$

For each $\mathcal{F}_{ij}$, we can choose an x_{ij} with $||x_{ij}|| = 1$ such that

$$||\mu_{\infty_{y_1}}(\mathcal{F}_{ij})|| < |\mu_{\infty_{y_1}}(\mathcal{F}_{ij}x_{ij})| + \frac{\epsilon}{k},$$

where k is a constant, by Theorem 7.3

$$\begin{aligned} HK\int_T ||f||d|\mu_{\infty_{y_1}}| &\le \sum_{ij} |\mu_{\infty_{y_1}}(\mathcal{F}_{ij})x_{ij}||x_{ij}||| + \epsilon \\ &= \sum_{i,j}\left|\mu_{\infty_{y_1}}(\mathcal{F}_{ij})x_{ij}\frac{||x_i||}{||f||_{\theta,M}}\right| ||f||_{\theta,M} + \epsilon \\ &\le |\mu_{\infty_{y_1}}|_{\theta,M}(T)||f||_{\theta,M} + \epsilon. \end{aligned}$$

So,

$$HK\int_T ||f||d|\mu_{\infty_{y_1}}| \le |\mu_{\infty_{y_1}}|_{\theta,M}(T)||f||_{\theta,M}.$$

Therefore

$$\mu_{\infty_{X,Y}}(f) \le ||\mu_\infty||_{\theta,M}(T)||f||_{\theta,M}.$$

From above we have

$$\mu_{\infty_{X,Y}}(f_n - f_m) \le ||\mu_\infty||_{\theta,M}(T)||f_n - f_m||_{\theta,M}.$$

This means $\{f_n\}$ is Cauchy sequence in $HK(\mu_{\infty_{X,Y}}, \mathcal{X})$. Thus $f \in HK(\mu_{\infty_{X,Y}}, \mathcal{X})$ as $f_n \to f$ is $\mu_{\infty_{X,Y}}$ a.e. and $\mu_{\infty_{X,Y}}(f_n - f) \to 0$ as $n \to \infty$. So, $\mu_{\infty_{X,Y}}(f) \le ||\mu_{\infty_{\theta,M}}||(T)||f||_{\theta,M}$ for $f \in \mathcal{H}^\theta(M, \mathcal{X})$ as this is true for each n. □

We will now prove $(\mathcal{H}^\theta, ||.||_{\mathcal{H}^\theta})$ is a solid Banach lattice with weak order unit. Recalling a partially ordered Banach space Z, which is also a vector lattice, is a Banach lattice if $||x|| \le ||y||$ for every $x, y \in Z$ with $|x| \le |y|$. A weak order unit of Z is a positive element $e \in Z$ such that if $x \in Z$ and $x\Lambda e = 0$, then $x = 0$. Let Z be a Banach lattice and $A \ne \emptyset \subset B \subset Z$. We say that A is solid in B if for each x, y with $x \in B,\ y \in A$ and $|x| \le |y|$, it is $x \in A$.

Theorem 7.5 $\left(\mathcal{H}^\theta, ||.||_{\mathcal{H}^\theta}\right)$ *is a solid Banach lattice with weak order unit.*

Proof It is clear that $(\mathcal{H}^\theta, ||.||_{\mathcal{H}^\theta})$ is a solid normed lattice with respect to the usual order.

By the Rybakov Theorem (see also [2] Theorem IX.2.2) there is $x_1' \in \mathcal{X}'$ with $||x_1'|| \le 1$ such that $\lambda = x_1'\mu_\infty$ is a control measure of μ_∞. If $f, g \in \mathcal{H}^\theta(|x'\mu_\infty|)$, $|f| \le |g|\ \lambda$,a.e, $k \in \mathbb{N}$ and $x' \in X'$ with $||x'|| \le 1$, then

$$\begin{aligned} &\inf\left\{k > 0 :\ \theta\left(\frac{|f|}{k}\right) d|x'\mu_\infty| \le 1\right\} \\ &\le \inf\left\{k > 0 :\ \theta\left(\frac{|g|}{k}\right) d|x'\mu_\infty| \le 1\right\}. \end{aligned}$$

That is $||f||_{\mathcal{H}^\theta} \leq ||g||_{\mathcal{H}^\theta}$. So, $\left(\mathcal{H}^\theta(|x'\mu_\infty|), ||.||_{\mathcal{H}^\theta}\right)$ is Banach lattice. Moreover, $\mathcal{H}^\theta(|x'\mu_\infty|)$ is easily a weak order unit. □

Theorem 7.6 *$(\mathcal{H}^\theta, ||.||_{\mathcal{H}^\theta})$ is a Köthe function spaces.*

7.3 Integral Representations

We discuss the integral representations of operators from $\mathcal{H}^\theta(M, \mathcal{X})$ into Y. We assume that M is relatively weakly compact in $Ca(\Sigma)$, then we can find a control measure $\lambda : \Sigma \to \mathbb{R}^+$ for M such that $\lambda \leq M$ and $M << \lambda$.

Proposition 7.3 *Let $f : T \to C$ be a λ-measurable function such that $fg \in HK(\lambda)$ for each $g \in \mathcal{H}^\theta(M)$, then* $\sup\{HK \int_T |fg| d\lambda : g \in \mathcal{H}^\theta(M),\ ||g||_{\theta,M} \leq 1\} < \infty$.

Proof If possible, let $\sup\{HK \int_T |fg| d\lambda : g \in \mathcal{H}^\theta(M),\ |g||_{\theta,M} \leq 1\} = \infty$, then for each n, there exists a sequence $\{g_n\} \subset \mathcal{H}^\theta(M)$ with $||g_n||_{\theta,M} \leq 1$ such that $HK \int_T |fg_n| d\lambda \geq n.2^n$. If $g_0(r) = \sum_{n=1}^{\infty} \frac{1}{2^n}|g_n(r)|$, then

$$
\begin{aligned}
||g_0||_{\theta,M} &\leq \frac{1}{2^n}||g_n||_{\theta,M} \\
&\leq \sum_{n=1}^{\infty} \frac{1}{2^n} \\
&= 1.
\end{aligned}
$$

So, $g_0 \in \mathcal{H}^\theta(M)$ and hence $fg_0 \in \mathcal{H}(\lambda)$. As

$$HK \int_T |fg_0| d\lambda \geq \frac{1}{2^n} HK \int_T |fg_n| d\lambda \geq n \ \textit{for each}\ n.$$

This shows fg_0 does not belong in $HK(\lambda)$. Hence

$$\sup\left\{HK \int_T |fg| d\lambda : g \in \mathcal{H}^\theta(M),\ ||g||_{\theta,M} \leq 1\right\} < \infty.$$

□

We will now discuss the existence of countable additive measure for a continuous linear operator from $\mathcal{H}^\theta(M, \mathcal{X})$ into Y.

Theorem 7.7 *Suppose $\Delta : \mathcal{H}^{\theta}(M, \mathcal{X}) \to Y$ is a continuous linear operator, then there exists a countable additive measure $\mu_{\infty} : \Sigma \to L(\mathcal{X}, Y)$ such that*

$$\Delta f = HK \int_T f d\mu_{\infty} \text{ for } f \in \mathcal{H}^{\theta}(M, \mathcal{X}).$$

Proof Suppose $\Delta : \mathcal{H}^{\theta}(M, \mathcal{X}) \to Y$ is a continuous linear operator. Let us define $\mu_{\infty} : \Sigma \to L(\mathcal{X}, Y)$ by $\mu_{\infty}(E)x = \Delta(x\chi_E)$. The linearity of $\mu_{\infty}(E)$, for each $E \in \Sigma$ is very obvious. Also,

$$\begin{aligned}||\mu_{\infty}(E)x|| &= ||\Delta x \chi_E|| \\ &\leq ||\Delta|| ||x\chi_E||_{\theta,M} \\ &\leq ||\Delta|| ||\chi_E||_{\theta,M}||x||.\end{aligned}$$

Therefore, Δ is a continuous linear operator. Now we need to show μ_{∞} is a countable additive measure. Since,

$$\begin{aligned}M_{\theta}(\chi_E)) &= \sup_{\mu_{\infty} \in M} HK \int_T \theta(\chi_E))d\mu_{\infty} \\ &= \theta(1)M(E).\end{aligned}$$

If $E_n \to \emptyset$, then $M_{\theta}(\chi_{E_n}) \to 0$. Using Proposition 7.1, $||\chi_{E_n}||_{\theta,M} \to 0$. Since μ_{∞} is finitely additive, the above limit shows that μ_{∞} is countable additive and $\Delta f = HK \int_T f d\mu_{\infty}$ for $f \in \mathcal{S}_{\mathcal{X}}(\Sigma)$. Now using Theorem 7.4, $\Delta f = HK \int_T f d\mu_{\infty}$ for $f \in \mathcal{H}^{\theta}(M, \mathcal{X})$. □

Theorem 7.8 *If $\Delta : \mathcal{H}^{\theta}(M, \mathcal{X}) \to Y$ defined by $\Delta f = HK \int_T f d\mu_{\infty}$ for $f \in \mathcal{H}^{\theta}(M, \mathcal{X})$ with $||\mu_{\infty}||_{\theta,M}(T) < \infty$, holds $||\Delta|| \leq \mu_{\infty\theta,M}(T) \leq 2||\Delta||$.*

We will now show that the converse of Theorem 7.8 is held for a countable additive measure.

Theorem 7.9 *If μ_{∞} is a countable additive measure and $\Delta : \mathcal{H}^{\theta}(M, \mathcal{X}) \to Y$ defined by $\Delta f = HK \int_T f d\mu_{\infty}$ for $f \in \mathcal{H}^{\theta}(M, \mathcal{X})$ with $||\mu_{\infty}||_{\theta,M}(T) < \infty$, then Δ is a continuous linear operator.*

Proof Suppose $\mu_{\infty} : \Sigma \to L(\mathcal{X}, Y)$ is a countable additive measure such that $||\mu_{\infty}||_{\theta,M}(T) < \infty$, then the mapping $\Delta : \mathcal{H}^{\theta}(M, \mathcal{X}) \to Y$ defined by $\Delta f = HK \int_T f d\mu_{\infty}$ for all $f \in \mathcal{H}^{\theta}(M, X)$ is linear. Also,

$$\begin{aligned}||\Delta f|| &= ||HK \int_T f d\mu_{\infty}|| \\ &= \sup_{||y_1|| \leq 1} \left| HK \int_T d|y_1\mu_{\infty}| \right|\end{aligned}$$

$$\leq \sup_{||y_1||\leq 1} HK \int_T ||f||d|y_1\mu_\infty|$$

$$= \mu_{\infty_{X,Y}}(f)$$

$$\leq ||\mu_\infty||_{\theta,M}(T)||f||_{\theta,M} \text{ using Proposition 7.2}$$

Hence $\Delta : \mathcal{H}^\theta(M, \mathcal{X}) \to Y$ is continuous. □

7.4 Geometrical Properties

In this section, we will discuss about the uniform convexity, reflexivity of $\mathcal{H}^\theta(M, \mathcal{X})$, and the Radon-Nikodym property of the Henstock-Orlicz spaces $\mathcal{H}^\theta(\mu_\infty)$. Before the discussion of uniform convexity and reflexivity of $\mathcal{H}^\theta(\mu_\infty)$, we state about Modular spaces. The Modular spaces can be defined as $B_{\theta,|x'\mu_\infty|}(f) = HK \int_T \theta(f)d|x'\mu_\infty|$. For complementary Young function θ and ϕ, we define

$$\mu_{\infty_\phi}(g) = HK \int_T \phi(|g|)d\mu_\infty, \; g \in S(\Sigma) \tag{7.6}$$

$$M_\phi(g) = \sup_{\mu_\infty \in M} \mu_{\infty_\phi}(g). \tag{7.7}$$

Definition 7.7 ([8]) The Modular $B_{\theta,|x'\mu_\infty|}$ is uniformly convex if for each $\epsilon > 0$ there exists a $p(\epsilon) \in (0, 1)$ such that $B_{\theta,|x'\mu_\infty|}(x) = B_{\theta,|x'\mu_\infty|}(y) = 1$, $B_{\theta,|x'\mu_\infty|}(x - y) \geq \epsilon$ implies $B_{\theta,|x'\mu_\infty|}(\frac{x+y}{2}) \leq 1 - p(\epsilon)$.

Proposition 7.4 *Let θ satisfy Δ_2 condition with μ_∞ being atomless and $\mu_\infty(T) < \infty$, then $\mathcal{H}^\theta(M, \mathcal{X})$ is locally uniformly convex if and only if pseudomodular $B_{\theta,|x'\mu_\infty|}(M, \mathcal{X})$ is locally uniformly convex.*

Proof Proof is similar to the technique Lemma 3 of [5]. □

Theorem 7.10 *$\mathcal{H}^\theta(M, \mathcal{X})$ is uniformly convex if and only if the Modular $B_{\theta,|x'\mu_\infty|}(M, \mathcal{X})$ is uniformly convex.*

Proof We can prove it by using Proposition 7.4 with a similar technique of Lemma 1 [6]. □

Theorem 7.11 *Assume M is nonatomic, then $\mathcal{H}^\theta(M, \mathcal{X})$ is reflexive.*

Proof Since uniform convex implies reflexivity, it is obvious. □

Recalling a Banach space $\mathcal{X}$ is said to have Radon-Nikodym property, shortly RNP, if given a finite measure space (T, Σ, μ_∞) and a vector measure $\mu_\infty : \Sigma \to X$ of finite variation and absolutely continuous with respect to μ_∞, then there exists a Bochner integrable function $g : \Sigma \to \mathcal{X}$ such that $\mu_\infty(E) = \int_E g d\mu_\infty$ for any $E \in \Sigma$. The fact that the Bochner integral is McShane integral (also see [3]), every McShane integral is Henstock integral (see page 158 of [4]). With this known fact we can define the Radon-Nikodym property as follows for our setting:

Definition 7.8 $\mathcal{H}^\theta(\Sigma)$ is said to have Radon-Nikodym property, shortly RNP, if given a finite measure space (T, Σ, μ_∞) and a vector measure $\mu_\infty : \Sigma \to \mathcal{H}^\theta$ of finite variation and absolutely continuous with respect to μ_∞, then there exists a Henstock integrable function $g : T \to \mathcal{H}^\theta$ such that $\mu_\infty(E) = HK \int_E g d\mu_\infty$ for any $E \in \Sigma$.

Definition 7.9 We say that $f \in \mathcal{X}$ has absolute norms if for every decreasing sequence $\{G_n\}$ of subset of $\mathcal{B}$ satisfying $\mu_\infty(G_n) \to 0$ we have $||f\chi_{G_n}|| \to 0$.

Theorem 7.12 *$(\mathcal{H}^\theta, H_\theta)$ has absolutely continuous norm.*

Proof Let $E_n = \{x \in \Sigma : n \le \theta(x) < n+1\}$, $n \in \mathbb{N}$. Then there exists a sequence of natural numbers $\{n_k\}_{k\in\mathbb{N}}$ such that $\mu_\infty(E_{n_k}) > 0$. Let $c_k > 0$ be such that $HK \int_{E_{n_k}} c_k^{\theta(x)} d|x'\mu_\infty| = 1$, $k \in \mathbb{N}$.

We have $f(x) = \sum\limits_{k=1}^{\infty} c_k \chi_{E_{n_k}}(x)$, $x \in \Omega$ and $E_j = \bigcup\limits_{k=j}^{\infty} E_{n_k}$. Then $E_n \to \phi$.

Now

$$\begin{aligned}||f|| &= \inf\left\{\alpha > 0 : HK \int_{E_{n_k}} \theta\left(\frac{f}{\alpha}\right) d|x'\mu_\infty| \le 1\right\}\\ &\le \inf\left\{\alpha > 1 : \sum_{k=1}^{\infty}\left(\frac{1}{\alpha}\right)^{n_k} \le 1\right\}\\ &\le 2.\end{aligned}$$

So, $f \in \mathcal{H}^\theta(\mu_\infty)$. Hence

$$||f\chi_{E_n}|| = \inf\left\{\alpha > 0 : HK \int_{E_{n_l}} \theta\left(\frac{c_l}{\alpha}\right) d|x'\mu_\infty| \le 1\right\}.$$

If $\mu_\infty(E_n) \to 0$ implies $||f\chi_{E_n}|| \to 0$. So, $\mathcal{H}^\theta(\mu_\infty)$ has absolutely continuous norm. □

Theorem 7.13 *Let $T = \bigcup\limits_{n=1}^{\infty} A_n$ be a union of measurable sets. If the subspace $\mathcal{H}_n^\theta = \{f \in \mathcal{H}^\theta : f = 0 \text{ on } T \setminus A_n\}$ of $\mathcal{H}^\theta$ have RNP, then $\mathcal{H}^\theta$ have RNP.*

Proof Let (T, Σ, μ_∞) be a finite measure space and let $\mu_\infty : \Sigma \to \mathcal{H}^\theta$ be a vector measure of finite variation which is absolutely continuous with respect to μ_∞. We define projections $\mathcal{P}_n : \mathcal{H}^\theta \to \mathcal{H}_n^\theta$ by $\mathcal{P}_n(f) = f\chi_{(A_n)}$ and $\mu_{\infty_n} = \mathcal{P}_n(\mu_\infty)$, then each μ_{∞_n} is an $\mathcal{H}_n^\theta$-valued vector measure of finite variation which is absolutely continuous with respect to μ_∞. With the fact that each space $\mathcal{H}_n^\theta$ has the RNP, there exists a Henstock integrable function $h_n : T \to \mathcal{H}_n^\theta$ satisfying $\mu_{\infty_n}(E) = \int_T h_n d\mu_\infty$ for each $E \in \Sigma$. Now for $E \in \Sigma$ and $n \in \mathbb{N}$ we have

$$HK \int_E || \sum_{k=1}^{n} h_k ||_{H_\theta} d\mu_\infty \leq |\mu_\infty|(E) \tag{7.8}$$

and so, for μ_∞-almost all $\alpha \in \Sigma$, there exists $h(\alpha) \in \mathcal{H}^\theta$ such that $h(\alpha) = \sum_{k=1}^{\infty} h_k(\alpha)$. Using (7.8) and Fatou's lemma, h is also Bochner integrable and so Henstock integrable. Lastly,

$$\begin{aligned}\mu_\infty(E) &= \lim_{n\to\infty} \sum_{k=1}^{n} \mu_{\infty_k}(E)\\ &= \lim_{n\to\infty} \sum_{k=1}^{n} HK \int_E h_n d\mu_\infty\\ &= HK \int_E g d\mu_\infty.\end{aligned}$$

Hence the proof. □

References

1. Alexiewicz, A.: Linear functionals on Denjoy-integrable functions. Collectanea Mathematica **1**, 289–293 (1948)
2. Duistermaat, J.J., Kolk, J.A.C.: Distribution: Theory and Application. Springer, Berlin Heidelberg, New York (2006)
3. Forkert, D.L.: The Banach space-valued integrals of Riemann, Mcshane, Henstock-Kurzweil and Bochner, Bachelor Thesis, Vienna University of Technology (2012)
4. Henstock, R.: The General Theory of Integration. Oxford Mathematical Monographs. Clarendon Press, Oxford (1991)
5. Kaminska, A.: The criteria for local Uniformly rotundity of Orlicz spaces. Studia Mathematica **79**(3), 201–215 (1984)
6. Kaminska, A.: On Uniform convexity of Orlicz spaces. Indagationes Mathematicae (Proceedings) **85**(1), 27–36 (1982)
7. Lindenstrauss, J., Tzafriri, L.: Classical Banach Spaces II, Function Spaces. A Series of Modern Surveys in Mathematics. Springer, Berlin-Heidelberg-New York (1979)
8. Luxemburg, W.A.J.: Banach Function Spaces, Thesis, Delft (1955)
9. Roy, S.K., Chakraborty, N.D.: Orlicz spaces for a family of measures. Analysis Mathematica **12**, 229–235 (1986)
10. Roy, S.K., Chakraborty, N.D.: On Orlicz spaces for a family of measures. J. Math. Anal. Appl. **145**(2), 485–503 (1990)

Chapter 8
Weak Henstock-Orlicz Spaces and Inclusion Properties

8.1 Introduction

In this chapter we focus on the completeness property of the weak H-Orlicz spaces without Δ_2 property which differs from the classical weak Orlicz spaces. Several convergence results are discussed. Inclusion properties of the weak H-Orlicz spaces are also presented in this chapter.

Definition 8.1 Quasi-Banach spaces: Let $\mathcal{X}$ be a vector space over $\mathbb{C}$. A function $||.|| : \mathcal{X} \to [0, \infty)$ is called a quasi norm if the following conditions are satisfied:

$$||f + g|| \leq c(||f|| + ||g||), \tag{8.1}$$

where $c(\geq 1)$ is a constant independent of $f, g \in \mathcal{X}$ and

$$||f|| \geq 0 \ (f \neq 0), \quad ||\lambda f|| = |\lambda| ||f|| \ (\lambda \in \mathbb{C}). \tag{8.2}$$

Then the pair $(\mathcal{X}, ||.||)$ is called a quasi-normed space . For $p < 1$, the functional $||.|| = ||.||_{\mathcal{L}^p}$ is not a norm for the Lebesgue spaces $\mathcal{L}^p$ space, but if $c^{\frac{1}{p-1}}$, then the inequality 8.1 is satisfied and

$$||f + g||^p \leq ||f||^p + ||g||^p. \tag{8.3}$$

The space $\mathcal{X}$ is endowed with the structure of a topological vector space by declaring "a neighborhood of zero" to mean "a set containing $\{f : ||f|| < \frac{1}{m}\}$ for some $m = 1, 2, \ldots$.." (The "ball" $\{f : ||f|| < 1\}$ need not be an open set. Therefore a quasi norm, in contrast to a p-norm, need not be continuous.) This topology is metrizable, if a p-norm $||.||$ is equivalent to the original quasi norm, then the formula $d(f, g) = ||f - g||^p$ defines a metric that induces the same topology. If $\mathcal{X}$ is complete in this

B. Hazarika, H. Kalita, *Theory of Henstock–Orlicz Spaces*, Mathematical Marvels: Texts and Monographs in the Spirit of CR Rao,
https://doi.org/10.1007/978-981-96-9548-5_8

topology, then it is called a quasi-Banach space. A p-Banach space is a complete p-normed space. If $\mathcal{X}$ is complete with respect to this topology, then it is called a quasi-Banach space. A p-Banach space is a complete p-normed space .

8.2 H-Orlicz Spaces with Locally Henstock-Kurzweil Integrals

In this section we discuss about H-Orlicz spaces with locally Henstock-Kurzweil integrals. H-Orlicz classes is an important part of H-Orlicz spaces. The H-Orlicz class can be stated as:

Definition 8.2 Let $\mathcal{H}^{-\theta}(\mu_\infty)$ be the set of all $f : \mathcal{K} \subset \mathbb{R}^n \to \mathbb{R}$ bounded measurable with compact support for $\mathcal{K} \subset \mathbb{R}^n$ such that $\int_{\mathcal{K}} \theta(|f|)d\mu_\infty$ is Henstock integrable, i.e.,

$$\mathcal{H}^{-\theta}(\mu_\infty) = \left\{ f : \int_{\mathcal{K}} \theta(|f|)d\mu_\infty \in HK(\mu_\infty) \right\},$$

whenever f is bounded measurable with compact support :

Theorem 8.1 *The space $\mathcal{H}^{-\theta}(\mu_\infty)$ is a linear space if and only if θ satisfies Δ_2-condition.*

Proof The proof is similar to [2, Theorem2.2]. □

Proposition 8.1 *For each $f \in \mathcal{H}^{\theta}(\mu_\infty)$, there is an $\alpha > 0$ such that*

$$\mathbb{B}_\theta = \left\{ \alpha f = g \in \mathcal{H}^{-\theta}(\mu_\infty) : HK \int_{\mathcal{K}} \theta(g)d\mu_\infty \le 1 \right\}$$

is a circled solid subset of $\mathcal{H}^{-\theta}(\mu_\infty)$.

Proof Let $f, g \in \mathcal{H}^{\theta}(\mu_\infty)$. Then there exist $\alpha, \beta > 0$ such that $\alpha f, \beta g \in \mathcal{H}^{-\theta}(\mu_\infty)$. Let $\nu = \min(\alpha, \beta)$. Then for $\nu > 0$ and using the fact of convexity and monotonicity of θ, we get

$$\begin{aligned} &HK \int_K \left(\frac{\nu}{2}(f+g)\right) d\mu_\infty \\ &\le \frac{1}{2}\left[HK \int_K \theta(\alpha f)d\mu_\infty) + HK \int_K \theta(\beta g)d\mu_\infty \right]. \end{aligned}$$

The right side is Henstock integrable. As $\frac{\nu}{2} > 0$, gives $f + g \in \mathcal{H}^{\theta}(\mu_\infty)$. In particular, with each f in $\mathcal{H}^{\theta}(\mu_\infty)$, $2f \in \mathcal{H}^{\theta}(\mu_\infty)$ and then $nf \in \mathcal{H}^{\theta}(\mu_\infty)$ for all integers $n > 1$, so that $\gamma f \in \mathcal{H}^{\theta}(\mu_\infty)$ for any scalar γ. Therefore the given set is solid and circled. To hold $\gamma f \in \mathcal{H}^{\theta}(\mu_{\infty})$ for some $\gamma > 0$. Let $a_n \to 0$ be arbitrary and set $\gamma_n = \min(\gamma, a_n)$. Then $\gamma_n \to 0$ and $\theta(\gamma_n f) \leq \theta(\gamma f)$ and $\theta(\gamma_n f) \to 0$ as θ is a continuous Young function. Dominated convergence theorem gives us $HK \int_K \theta(\gamma_n f) \to 0$ so that for some n_0, we have $HK \int_K \theta(\gamma_{n_o} f) d\mu_{\infty} \leq 1$. Thus $\gamma_{n_0} f \in \mathbb{B}_{\theta}$. □

To construct $\mathcal{H}^{\theta}(\mathbb{R}^n)$ in our setting recall the known facts that $HK(\mathbb{R}^n) \subset HK_{loc}(\mathbb{R}^n)$. For a Young function θ, we can state our H-Orlicz space as follows:

$$\mathcal{H}^{\theta}(\mathbb{R}^n) = \{f \in HK_{loc} : ||f||_{H^{\theta}} < +\infty\},$$

where $||f||_{H^{\theta}(\mathbb{R}^n)}$ is defined as follows:

$$H_{\theta}(f) = \inf\left\{a > 0 : HK \int_{\mathcal{K}} \theta\left(\frac{f}{a}\right) d\mu_{\infty} \leq 1\right\},$$

where $\mathcal{K}$ is a compact subset of $\mathbb{R}^n$. The following results are held with this settings.

Proposition 8.2 *$\mathcal{H}^{\theta}(\mu_{\infty})$ is a linear space.*

Proof Let $f_i \in \mathcal{H}^{\theta}(\mu_{\infty})$, $i = 1, 2$. Then there exists α_i such that by definition $\alpha_i f_i \in \mathcal{H}^{-\theta}(\mu_{\infty})$.

Let $\alpha = \min\{\alpha_1, \alpha_2\}$. Then for $\alpha > 0$.

$$HK \int_{\mathcal{K}} \theta\left(\frac{\alpha}{2}(f_1 + f_2)\right) d\mu_{\infty}$$
$$\leq \frac{1}{2}\left[HK \int_{\mathcal{K}} \theta(\alpha_1 f_1) d\mu_{\infty} + HK \int_{\mathcal{K}} \theta(\alpha_2 f_2) d\mu_{\infty}\right],$$

where $\mathcal{K}$ is a compact subset of $\mathbb{R}^n$. The right-hand side is Henstock integrable. So, $f_1 + f_2 \in \mathcal{H}^{\theta}(\mu_{\infty})$ as $\frac{\alpha}{2} > 0$.

If $f = f_1 = f_2$ implies $2f \in \mathcal{H}^{\theta}(\mu_{\infty})$. So, $nf \in \mathcal{H}^{\theta}(\mu_{\infty})$ for all integer $n > 1$. Therefore $\beta f \in \mathcal{H}^{\theta}(\mu_{\infty})$ for any scalar β. Hence $\mathcal{H}^{\theta}(\mu_{\infty})$ is a linear space. □

Theorem 8.2 *The space $(\mathcal{H}^{\theta}(\mathbb{R}^n), ||.||_{H_{\theta}})$ is normed linear space.*

Proof

(a) If $f = 0$ a.e., then $H_{\theta}(f) = 0$.

Conversely, consider $H_{\theta}(f) = 0$.

Let $|f| > 0$, on a set of positive measure. If possible then there exists a number $\delta > 0$ such that for a compact subset $\mathcal{K} = \{p : |f(p)| \geq \delta\}$ of $\mathbb{R}^n$, satisfies $\mu_{\infty}(\mathcal{K}) > 0$.

Now

$$\begin{aligned}\theta(n\delta)\mu_\infty(A) &= \int_K \theta(n\delta)d\mu_\infty \\ &\le \int_{\mathcal{K}} \theta(nf)d\mu_\infty \\ &\le \int_{\mathcal{K}} \theta(nf)d\mu_\infty,\ n \ge 1.\end{aligned}$$

Since $\mu_\infty(\mathcal{K}) > 0$ and $\theta(n\delta)d\mu_\infty \to \infty$ as $n \to \infty$. This is not possible so, $\mu_\infty(\mathcal{K}) = 0$. Thus $f = 0$ a.e. Therefore $H_\theta(f) = 0$ if and only if $f = 0$ a.e.

(b) $H_\theta(\alpha f) = |\alpha| H_\theta(f),\ \alpha \in \mathbb{R}$ or $\alpha \in \mathbb{C}$.
Let $\alpha \neq 0$.

$$\begin{aligned}H_\theta(\alpha f) &= \inf\left\{\alpha_1 > 0 : HK\int_{\mathcal{K}} \theta\left(\frac{\alpha f}{\alpha_1}\right) d\mu_\infty \le 1\right\} \\ &= |\alpha| \inf\left\{\frac{\alpha_1}{|\alpha|} > 0 : HK\int_{\mathcal{K}} \theta\left(\frac{f}{\frac{\alpha_1}{|\alpha|}}\right) d\mu_\infty \le 1\right\} \\ &= |\alpha| H_\theta(f).\end{aligned}$$

(c) $H_\theta(f+g) \le H_\theta(f) + H_\theta(g)$.
Let $f, g \in \mathcal{H}^\theta(\mu_\infty)$, $a > H_\theta(f)$ and $b > H_\theta(g)$. Then $0 < a < \infty$ and $0 < b < \infty$, and suppose $c = a+b > 0$. Since $f+g \in \mathcal{H}^\theta(\mu_\infty)$, $H_\theta(f+g) < \infty$. Now

$$\begin{aligned}HK\int_{\mathcal{K}} \theta\left(\frac{f+g}{c}\right) d\mu_\infty &= HK\int_{\mathcal{K}} \theta\left[\frac{f}{a}.\frac{a}{c} + \frac{g}{b}.\frac{b}{c}\right] d\mu_\infty \\ &\le \frac{a}{c} HK\int_{\mathcal{K}} \theta\left(\frac{f}{a}\right) d\mu_\infty \\ &+ \frac{b}{c} HK\int_{\mathcal{K}} \theta\left(\frac{g}{b}\right) d\mu_\infty \\ &\le \frac{a}{c} + \frac{b}{c} \\ &\le 1.\end{aligned}$$

By Proposition 8.1, we have $\frac{1}{c}(f+g) \in \mathbb{B}_\theta$ so that $H_\theta(f+g) \le c = a+b$. If $a \to H_\theta(f)$ and $b \to H_\theta(g)$, we get the triangle inequality $H_\theta(f+g) \le H_\theta(f) + H_\theta(g)$. Hence $(\mathcal{H}^\theta, ||.||_{H_\theta})$ is a normed linear space.

□

Lemma 8.1 *For any bounded measurable function* g *on* $\mathbb{R}^n$, $||g||_{H_\theta} = 0$ *if and only if* $g = 0$ *a.e.*

Lemma 8.2 *If* $0 < ||f||_{H_\theta} < \infty$, *then* $HK \int_{I \subseteq \mathbb{R}^n} \theta\left(\frac{f}{||f||_{H_\theta}}\right) \leq A$, *where* $A = \lim S\left(\theta\left(\frac{f}{||f||_{H_\theta}}\right), D_k\right)$, D_k *is a sequence of tagged partition.*

Theorem 8.3 *For all* $f \in \mathcal{H}^\theta(\mathbb{R}^n)$, $g \in \mathcal{H}^\Phi(\mathbb{R}^n)$, *where* Φ *is complementary function of* θ, *then*

$$HK \int_{\mathbb{R}^n} |fg| d\mu_\infty \leq ||f||_{H_\theta} ||g||_{H_\Phi}.$$

Theorem 8.4 *The classical Orlicz space is a dense subspace of H-Orlicz space as continuous dense embeddings. That is* $\mathcal{L}^\theta(\mathbb{R}^n) \subseteq \mathcal{H}^\theta(\mathbb{R}^n)$ *as continuous dense embeddings.*

Proof Let $f \in \mathcal{L}^\theta(\mathbb{R}^n)$. Then $f \in L_{loc}$ with $||f||_{L^\theta(\mathbb{R}^n)} < \infty$. Then for some $\lambda > 0$, and a compact $\mathcal{K} \subset \mathbb{R}^n$, we have

$$\begin{aligned}
&\inf\left\{\alpha > 0 : HK \int_{\mathcal{K}} \theta\left(\frac{f(x)}{\alpha}\right) d\mu_\infty\right\} \\
&\leq \inf\left\{\alpha > 0 : L \int_{\mathbb{R}^n} \theta\left(\frac{|f(x)|}{\alpha}\right) d\mu_\infty\right\} \\
&\leq 1.
\end{aligned}$$

So, for some $\alpha > 0$, $\inf\left\{\alpha > 0 : L \int_{\mathbb{R}^n} \theta\left(\frac{|f(x)|}{\alpha}\right) d\mu_\infty \leq 1\right\}$ implies for some $\alpha > 0$, we get the following,

$$\inf\left\{\alpha > 0 : HK \int_{\mathbb{R}^n} \theta\left(\frac{f(x)}{\alpha}\right) d\mu_\infty \leq 1\right\}.$$

Hence $f \in \mathcal{H}^\theta(\mathbb{R}^n)$ with $||f||_{H^\theta(\mathbb{R}^n)} \leq ||f||_L^\theta(\mathbb{R}^n)$. Hence the proof. □

Remark 8.1 If in $f \in \mathcal{L}^\theta(\mathbb{R}^n)$, then all the theorems that are true in $\mathcal{L}^\theta(\mathbb{R}^n)$ are also true for $\mathcal{H}^\theta(\mathbb{R}^n)$.

8.3 Weak-Henstock-Orlicz Spaces

In this section we discuss about the weak Henstock-Orlicz spaces and its basic properties. The classical weak Orlicz space is not naturally Banach spaces. We discuss the weak Henstock-Orlicz space is naturally a Banach space. We consider all functions f are μ_∞ measurable now onward. For a measurable set $\mathcal{B} \subset \mathbb{R}^n$, we denote the characteristic function of $\mathcal{B}$ by $\chi_{\mathcal{B}}$. For a measurable set $\mathcal{B}$, a measurable function f and $t > 0$, let

$$\mu_\infty(\mathcal{B}, f, t) = \{x \in \mathcal{B} : \mu_\infty(f(x)) > t\}.$$

If $\mathcal{B} = \mathbb{R}^n$, we denote as $\mu_\infty(f, t)$.

Definition 8.3 Let f be μ_∞ measurable functions. Then weak-H-Orlicz space is defined as:

$$\mathcal{H}^\theta_w(\mathbb{R}^n) = \left\{ f \in HK_{loc}(\mathbb{R}^n) : \; ||f||_{H^\theta_w} < +\infty \right\},$$

where $||f||^\theta_w = \inf \left\{ \exists\, \alpha > 0 : \; \sup_{t>0} \theta(t) \mu_\infty\left(\frac{f}{\alpha}, t\right) \leq 1 \right\}$.

We consider another μ_∞-measurable set as

$$B^\theta_w(\mathbb{R}^n) = \left\{ f \in HK_{loc}(\mathbb{R}^n) : \; \forall\, \alpha > 0 : \sup_{t>0} \theta(t) \mu_\infty\left(\frac{f}{\alpha}, t\right) < \infty \right\}.$$

It is observed that $B^\theta_w(\mathbb{R}^n) \subset \mathcal{H}^\theta_w(\mathbb{R}^n)$ and $B^\theta_w(\mathbb{R}^n)$ is a linear space. If $f = g$ are μ_∞ a.e. in $\mathcal{H}^\theta_w(\mathbb{R}^n)$, then we call $f = g$ in $\mathcal{H}^\theta_w(\mathbb{R}^n)$. We define the module of function f as

$$\Upsilon_\theta(f) = \sup_{t>0} \theta(t) \mu_\infty\left(\frac{f}{\alpha}, t\right).$$

Lemma 8.3

1. *If $f \leq g$ are μ_∞-a.e. and $g \in \mathcal{H}^\theta_w$, then $f \in \mathcal{H}^\theta_w$.*
2. $\sup_{t>0} \theta\left(\frac{t}{||f||_{H^\theta_w}}\right) \mu_\infty(f, t) \leq 1.$
3. *If $\theta \in \Delta_2$ and $||f||_{H^\theta_w(\mathbb{R}^n)} \leq 1$, then $\Upsilon_\theta(f) \leq ||f||_{H^\theta_w(\mathbb{R}^n)}$.*

Proof The proof of (1) is very straightforward as $f \leq g$ are μ_∞-a.e. means $|f| \leq |g|$ are μ_∞-a.e.

For (2) from the definition of $||f||_{H^\theta_w(\mathbb{R}^n)}$ there exists $C_k \searrow ||f||_{H^\theta_w(\mathbb{R}^n)}$ such that $\theta\left(\frac{f}{C_k}\right) \mu_\infty(f, t) \leq 1$, $\forall\, t > 0$. When $k \to \infty$, $\sup_{t>0} \theta\left(\frac{t}{||f||_{H^\theta_w}}\right) \mu_\infty(f, t) \leq 1$. □

Theorem 8.5 *If $\theta \in \Delta_2$, then $\left(\mathcal{H}^\theta_w, ||.||_{H^\theta_w}\right)$ is a quasi-Banach space.*

Proof The proof of theorem follows from of [5, Lemma1.1(4)]. □

Theorem 8.6 *For $f, g \in \mathcal{H}^\theta_w$, the following inequalities hold:*

$$||f + g||_{H^\theta_w} \leq ||f||_{H^\theta_w} + ||g||_{H^\theta_w}.$$

Lemma 8.4 *If $\theta \in \Delta_2$, then the following hold:*

1. $\mathcal{H}_w^\theta(\mathbb{R}^n) = B_w^\theta(\mathbb{R}^n)$.
2. *If $f_n \in \mathcal{H}_w^\theta(\mathbb{R}^n)$, then $||f_n - f||_{H_w^\theta(\mathbb{R}^n)} \to 0$ if $\Upsilon_\theta(f_n - f) \to 0$.*
3. *If $f_n \in \mathcal{H}_w^\theta(\mathbb{R}^n)$, then $\sup\left\{||f_n||_{H_w^\theta(\mathbb{R}^n)}\right\} \leq k$ if and only if $\sup\{\Upsilon_\theta(f_n)\} \leq k$, where $k > 0$.*

Proof For (1) Let $\theta \in \Delta_2$ and $f \in \mathcal{H}_w^\theta(\mathbb{R}^n)$, then there exists $\alpha > 0$ such that

$$\sup_{t>0} \theta\left(\frac{t}{\alpha}\right) \mu_\infty(f, t) \leq 1.$$

Hence $\theta\left(\frac{t}{a}\right) \leq \theta\left(\frac{t}{\alpha}\right)$ when $a \geq \alpha$. As $\theta \in \Delta_2$ so, $\theta\left(\frac{t}{a}\right) \leq c\theta\left(\frac{t}{\alpha}\right)$ when $a < \alpha$. So,

$$\sup_{t>0} \theta\left(\frac{t}{a}\right) \mu_\infty(f, t) \leq 1 \ \textit{for all} \ a > 0.$$

Hence $f \in B_w^\theta(\mathbb{R}^n)$ and hence the complete proof.

The proof of (2) follows the similar technique as [5, Corollary 2.1] and the proof of (3) follows the similar technique as [5, Theorem 2.1(4)]. □

The weak Orlicz space is not naturally complete. The weak Orlicz space is complete if the Young function $\theta \in \Delta_2$. We observe that the weak H-Orlicz is complete if the Young function $\theta \in \Delta_2$. We discuss now that the weak H-Orlicz space is naturally complete.

Lemma 8.5 *Let $f_n, \in \mathcal{H}_w^\theta(\mathbb{R}^n)$, then the following are true*

1. *If $||f_n - f||_{H_w^\theta} \to 0$, then $f_n \to f$ (convergence in measure).*
2. *If $0 \leq \inf(f_n) \to f$ is μ_∞-a.e., then $||f_n||_{H_w^\theta} \to ||f||_{H_w^\theta}$.*

Proof (1) From the definition of the norm of the weak H-Orlicz space, we have

$$||f_n - f||_{H_w^\theta(\mathbb{R}^n)} = \inf\left\{\alpha > 0 : \sup_{t>0} \theta\left(\frac{f_n - f}{\alpha}\right)\mu_\infty(f_n - f, t)\right\} \to 0$$

gives, $\theta\left(\frac{f_n - f}{\alpha}\right) \mu_\infty(f_n - f, t) \to 0$. This implies $\mu_\infty(f_n - f) \to 0$. Therefore $f_n \to f$ in the measure μ_∞.

For (2): Let $\inf \mu_\infty(f_n) = \mu_\infty(f)$ μ_∞-a.e. Then clearly $f_n \to f$ in the measure μ_∞. This gives for all $\epsilon > 0$ there exists a number $n_0 \in \mathbb{N}$ such that

$$\mu_\infty(f_n - f) \to 0 \ \textit{for} \ n > n_0.$$

Using the property for a Young function θ, we can write

$$\theta\left(\frac{f_n - f}{\alpha}\right)\mu_\infty(f_n - f, t) \to 0$$

$$\Rightarrow \sup\theta\left(\frac{f_n - f}{\alpha}\right)\mu_\infty(f_n - f, t) \to 0.$$

That is $\inf\left\{\alpha > 0 : \ \sup\theta\left(\frac{f_n - f}{\alpha}\right)\mu_\infty(f_n - f, t)\right\} < \epsilon$. As ϵ is an arbitrary, so

$$||f_n - f||_{H_w^\theta(\mathbb{R}^n)} \to 0.$$

□

Theorem 8.7 $\left(\mathcal{H}_w^\theta(\mathbb{R}^n), ||.||_w^\theta(\mathbb{R}^n)\right)$ *is a Banach space.*

Proof Let $f_n \in \mathcal{H}_w^\theta(\mathbb{R}^n)$ be a Cauchy sequence. Then there exists a natural number n_0 such that $\lim\limits_{n,m\to\infty} ||f_n - f_m||_{H_w^\theta(\mathbb{R}^n)} = 0$ for all $n, m \geq n_0$. We know $\lim\limits_{n,m\to n_0} \mu_\infty(f_n - f_m, t) = 0$ for all $t > 0$ implies $f_n \to f$ as μ_∞ measurable. By Riesz's theorem, there is a subsequence $f_{n_k} \to f$, μ_∞-a.e and $\mu_\infty(f_{n_k} - f_{n_s}) \to \mu_\infty(f_{n_k} - f)$ as μ_∞-a.e. Let $n_k \geq n_0$ and $f_{n_s} \to \infty$. Using (2) of the Lemma 8.5, $||f_{n_k} - f||_{H_w^\theta(\mathbb{R}^n)} < \epsilon$ and $\lim\limits_{n_k\to\infty} ||f_{n_k} - f||_{H_w^\theta(\mathbb{R}^n)} = 0$. Now from Theorem 8.6, we can find the following:

$$||f||_{H_w^\theta(\mathbb{R}^n)} \leq ||f_{n_k} - f||_{H_w^\theta(\mathbb{R}^n)} + ||f_{n_k}||_{H_w^\theta(\mathbb{R}^n)}.$$

This implies $||f||_{H_w^\theta(\mathbb{R}^n)} \leq ||f_{n_k}||_{H_w^\theta(\mathbb{R}^n)}$. Using the first part of Lemma(8.3) we can conclude that $f \in \mathcal{H}_w^\theta(\mathbb{R}^n)$. So, $f_n \to f$ in $\mathcal{H}_w^\theta(\mathbb{R}^n)$. □

Theorem 8.8 $\mathcal{L}_w^\theta(\mathbb{R}^n) \subseteq \mathcal{H}_w^\theta(\mathbb{R}^n)$ *as continuous dense embeddings.*

Proof Let $f \in \mathcal{L}_w^\theta(\mathbb{R}^n)$, then $f \in \mathcal{L}_{loc}^1(\mathbb{R}^n)$ with $||f||_{H_w^\theta(\mathbb{R}^n)} < \infty$. This is very obvious that $f \in HK_{loc}$. Now we find $||f||_{H_w^\theta(\mathbb{R}^n)} < \infty$. From the fact that for some $\alpha > 0$,

$$||f||_{H_w^\theta(\mathbb{R}^n)} = \inf\left\{\alpha > 0 : \ \sup_{t>0}\theta(t)\mu_\infty\left(\frac{f}{\alpha}, t\right) \leq 1\right\}.$$

We get $\sup\limits_{t>0}\theta(t)\mu_\infty(\frac{f}{\alpha}, t) \leq 1$ also true when $f \in \mathcal{L}_{loc}^1(\mathbb{R}^n)$. That is $||f||_{H_w^\theta(\mathbb{R}^n)} \leq ||f||_{L_w^\theta(\mathbb{R}^n)}$. Hence $\mathcal{L}_w^\theta(\mathbb{R}^n) \subseteq \mathcal{H}_w^\theta(\mathbb{R}^n)$ as continuous dense embeddings. □

Remark 8.2 The Δ_2 condition of the Young function θ is not necessary to prove the Bounded convergence theorem, Control convergence theorem, Fatou-type convergence theorem, Levi-type convergence theorem, and Vitali-type convergence theorem that are proved in [7, Section 3] in the weak H-Orlicz space.

8.4 Inclusion Property of Weak H-Orlicz Spaces

In this section we discuss the inclusion property of weak H-Orlicz spaces. Before that we find inclusion between H-Orlicz spaces and weak H-Orlicz spaces in the following theorem as follows:

Theorem 8.9 *Let θ be a Young function, then $\mathcal{H}^{\theta}(\mathbb{R}^n) \subset \mathcal{H}_w^{\theta}(\mathbb{R}^n)$ for every $f \in \mathcal{H}^{\theta}(\mathbb{R}^n)$ with $||f||_{H_w^{\theta}(\mathbb{R}^n)} \leq ||f||_{H^{\theta}}(\mathbb{R}^n)$.*

Proof Let $f \in \mathcal{H}^{\theta}(\mathbb{R}^n)$. We need to prove $f \in \mathcal{H}_w^{\theta}(\mathbb{R}^n)$. Let

$$\mathfrak{A}_{\theta,w} = \left\{\alpha > 0 : \sup_{t>0} \theta(t)\mu_{\infty}\left(\frac{f}{\alpha}, t\right) \leq 1\right\}$$

and

$$\mathfrak{B}_{\theta,w} = \left\{\alpha > 0 : HK \int_{\mathbb{R}^n} \theta\left(\frac{f}{\alpha}\right) d\mu_{\infty} \leq 1\right\}.$$

Then clearly, $||f||_{H_w^{\theta}(\mathbb{R}^n)} = \inf \mathfrak{A}_{\theta,w}$ and $||f||_{H^{\theta}(\mathbb{R}^n)} = \inf \mathfrak{B}_{\theta,w}$. Now for any $\beta \in \mathfrak{B}_{\theta,w}$ and $t > 0$ we have

$$\begin{aligned}
\theta(t)\mu_{\infty}\left(\frac{f}{\beta}, t\right) &\leq HK \int_{\{x\in\mathbb{R}^n:\ \mu_{\infty}(\frac{f}{\beta},t)\}} \theta\left(\frac{f}{\beta}\right) d\mu_{\infty} \\
&\leq HK \int_{\mathbb{R}^n} \theta\left(\frac{f}{\beta}\right) d\mu_{\infty} \\
&\leq 1.
\end{aligned}$$

As $t > 0$ is arbitrary, $\sup_{t>0} \theta(t)\mu_{\infty}\left(\frac{f}{\beta}, t\right) \leq 1$ and $\mathfrak{B}_{\theta,w} \leq \mathfrak{A}_{\theta,w}$. Hence $f \in \mathcal{H}_w^{\theta}(\mathbb{R}^n)$ with $||f||_{H_w^{\theta}(\mathbb{R}^n)} \leq ||f||_{H^{\theta}}(\mathbb{R}^n)$. □

Lemma 8.6 *Let θ be a Young function $a \in \mathbb{R}^n$ and $r > 0$ be arbitrary, then*

$$||\chi_{B(a,r)}||_{H_w^{\theta}(\mathbb{R}^n)} = \frac{1}{\theta^{-1}\left(\dfrac{1}{\mu_{\infty}(B(a,r))}\right)},$$

where $\mu_{\infty}(B(a,r))$ is the volume of open ball $B(a,r)$.

Proof Theorem 8.9 gives $||f||_{H_w^\theta(\mathbb{R}^n} \leq ||f||_{H^\theta}(\mathbb{R}^n)$. Now,

$$\begin{aligned}
||\chi_{B(a,r)}||_{H^\theta(\mathbb{R}^n)} &= HK \int_{\mathbb{R}^n} \theta\left(\chi_{B(a,r)}\theta^{-1}\left(\frac{1}{\mu_\infty(B(a,r))}\right)\right) d\mu_\infty \\
&= HK \int_{B(a,r)} \theta\left(\chi_{B(a,r)}\theta^{-1}\left(\frac{1}{\mu_\infty(B(a,r))}\right)\right) d\mu_\infty \\
&\leq HK \int_{B(a,r)} \frac{1}{\mu_\infty B(a,r)} d\mu_\infty \\
&= \frac{1}{\mu_\infty(B(a,r))} HK \int_{B(a,r)} d\mu_\infty \\
&= 1.
\end{aligned}$$

By the definition of $||.||_{H^\theta(\mathbb{R}^n)}$, we get

$$||\chi_{B(a,r)}||_{H_w^\theta(\mathbb{R}^n)} \leq \frac{1}{\theta^{-1}\left(\frac{1}{\mu_\infty(B(r,a))}\right)}.$$

Now we need to prove

$$||\chi_{B(a,r)}||_{H_w^\theta(\mathbb{R}^n)} \geq \frac{1}{\theta^{-1}\left(\frac{1}{\mu_\infty(B(r,a))}\right)}.$$

If possible let $||\chi_{B(a,r)}||_{H_w^\theta(\mathbb{R}^n)} < \frac{1}{\theta^{-1}\left(\frac{1}{\mu_\infty(B(r,a))}\right)}$. Then by the definition of $||.||_{H_w^\theta(\mathbb{R}^n)}$, we find $||\chi_{B(a,r)}||_{H_w^\theta(\mathbb{R}^n)} \leq 1$. This contradicts our assumption and hence

$$||\chi_{B(a,r)}||_{H_w^\theta(\mathbb{R}^n)} = \frac{1}{\theta^{-1}\left(\frac{1}{\mu_\infty(B(r,a))}\right)}.$$

□

Theorem 8.10 *Let θ, ϕ be Young functions, then the following statements are equivalent:*

1. $\theta(t) \leq \phi(Ct)$ *for every* $t > 0$.
2. $\mathcal{H}_w^\phi(\mathbb{R}^n) \subseteq \mathcal{H}_w^\theta(\mathbb{R}^n)$.
3. *For every* $f \in \mathcal{H}_w^\theta(\mathbb{R}^n)$, *we have* $||f||_{H_w^\theta(\mathbb{R}^n)} \leq C||f||_{H_w^\phi(\mathbb{R}^n)}$.

Proof The proof is similar to [6, Theorem 3.3] □

Now we state a necessary and sufficient condition for the inclusion properties of weak H-Orlicz spaces generated by concave function presented in the following theorem.

Theorem 8.11 *Let $\theta_1, \theta_2 \in \mathfrak{F}$. The following are equivalent:*

1. $\theta_1 \Delta \theta_2$.
2. $\mathcal{H}_w^{\theta_2}(\mathbb{R}^n) \subseteq \mathcal{H}_w^{\theta_1}(\mathbb{R}^n)$.
3. *There exists a constant $C > 0$ such that $||f||_{H_w^{\theta_1}}(\mathbb{R}^n) \leq C||f||_{H_w^{\theta_2}}(\mathbb{R}^n)$ for every $f \in \mathcal{H}_w^{\theta_2}(\mathbb{R}^n)$.*

Proof For (1) $\Rightarrow$ (2): Let $f \in \mathcal{H}_w^{\theta_2}(\mathbb{R}^n)$,

$$\mathbb{A}_{\theta_1} = \left\{\alpha > 0 : \sup_{t>0} \theta_1(t)\mu_\infty\left(\frac{f}{\alpha}, t\right) \leq 1\right\}$$

and

$$\begin{aligned}\mathbb{A}_{\theta_2} &= \left\{\alpha > 0 : \sup_{t>0} \theta_2(Ct)\mu_\infty\left(\frac{f}{\alpha}, t\right) \leq 1\right\} \\ &= \left\{\alpha > 0 : \sup_{t>0} \theta_2(p)\mu_\infty\left(\frac{Cf}{\alpha}, p\right) \leq 1\right\}\end{aligned}$$

for $p = Ct$. If $t > 0$ and $\alpha \in \mathbb{A}_{\theta_2}$, then

$$\begin{aligned}\theta_1(t)\mu_\infty\left(\frac{f}{\alpha}, t\right) &\leq \theta_2(Ct)\mu_\infty\left(\frac{f}{\alpha}, t\right) \\ &= \theta_2(p)\mu_\infty\left(\frac{cf}{\alpha}, t\right) \\ &\leq 1.\end{aligned}$$

So,

$$\begin{aligned}||f||_{H_w^{\theta_1}(\mathbb{R}^n)} &= \inf \mathbb{A}_{\theta_1} \\ &\leq \inf \mathbb{A}_{\theta_2} \\ &= C||f||_{H_w^{\theta_2}(\mathbb{R}^n)}.\end{aligned}$$

Therefore $\mathcal{H}_w^{\theta_2}(\mathbb{R}^n) \subseteq \mathcal{H}_w^{\theta_1}(\mathbb{R}^n)$.

For (2) $\Rightarrow$ (3): Since $\left(\mathcal{H}_w^{\theta_2}(\mathbb{R}^n), \mathcal{H}_w^{\theta_1}(\mathbb{R}^n)\right)$ are pair of Banach spaces, then by [3, Lemma3.3]) we get the conclusion.

For (3) $\Rightarrow$ (1): From Lemma(8.6)

$$\begin{aligned}\frac{1}{\theta_1^{-1}\left(\frac{1}{\mu_\infty(B(r,a))}\right)} &= ||\chi_{B(a,r)}||_{H_w^{\theta}(\mathbb{R}^n)}\\ &\leq C||\chi_{B(a,r)}||_{H_w^{\theta_2}(\mathbb{R}^n)}\\ &= C\frac{1}{\theta_2^{-1}\left(\frac{1}{\mu_\infty(B(r,a))}\right)}.\end{aligned}$$

Therefore for any $a \in \mathbb{R}^n$ and $r > 0$ we find

$$C\theta_1^{-1}\left(\frac{1}{\mu_\infty(B(a,r))}\right) \geq \theta_2^{-1}\left(\frac{1}{\mu_\infty(B(a,r))}\right).$$

Using [6, Lemma1.1(4)] we get $\theta_1\left(\frac{1}{\mu_\infty(B(a,r))}\right) \leq \theta_2\left(\frac{C}{\mu_\infty(B(a,r))}\right)$, □

since $r > 0$ is arbitrary, assuming $t = \frac{1}{\mu_\infty((B(a,r))}$, we get the conclusion $\theta_1(t) \leq \theta_2(Ct)$ and hence $\theta_1 \, \Delta \, \theta_2$.

References

1. Alexiewicz, A.: Linear functionals on Denjoy-integrable functions. Collectanea Mathematica **1**, 289–293 (1948)
2. Hazarika, B., Kalita, H.: Henstock-Orlicz spaces and its dense spaces. Asian Eur. J. Math. **14**(7), 1–17 (2021)
3. Krein, S.G., Petunia, Yu.I., Semënov, E.M.: Interpolation of Linear Operators, Translation of Mathematical Monograph 54. American Mathematical Society, Providence (1982)
4. Lindenstrauss, J., Tzafriri, L.: Classical Banach Spaces II, Function Spaces. A Series of Modern Surveys in Mathematics. Springer, Berlin-Heidelberg-New York (1979)
5. Liu, P., Wang, M.: Weak Orlicz spaces: Some basic properties and their applications to harmonic analysis. Sci. China Math. **56**(4), 789–802 (2013)
6. Masta, A., Gunawan, A.H., Budhi, W.S.: Inclusion properties of Orlicz and weak Orlicz spaces. J. Math. Fund. Sci. **48**(3), 193–203 (2016)
7. Ning, L. Yonggang, Y.: Weak Orlicz space and its convergence theorems. Acta Mathematica Scientia **30B**(5), 1492–1500 (2010)

Chapter 9
Countable Additivity of Henstock-Dunford Integrable Function and Orlicz Spaces

9.1 Introduction

Let $\mathcal{I}_0$ be a compact interval in $\mathbb{R}^m$ (or $\mathbb{R}^1$) and $\mathcal{E} \subset \mathbb{R}^m$ (or $\mathbb{R}$) a measurable subset. Also, $\mu_\infty(\mathcal{E})$ stands for the Lebesgue measure of $\mathcal{E}$. The Lebesgue integral of a function g over the set $\mathcal{E}$ will be denoted by $L \int_{\mathcal{E}} g$. Throughout the chapter, $\mathcal{X}$ is a real Banach space with norm $||.||$ and $\mathcal{X}^*$ is its dual. $B^*_{\mathcal{X}} = \{y^* \in \mathcal{X} : ||y^*|| \leq 1\}$ is the closed unit ball in $\mathcal{X}^*$. In [16] we found that $HK(\mu_\infty)$ is equivalent to $\mathcal{L}^1(\mu_\infty)$, where $HK(\mu_\infty)$ is Henstock-Kurzweil integrable function space (in brief Henstock integrable function space) and $\mathcal{L}^1(.)$ is classical Lebesgue space.

9.2 Weakly Henstock-Kurzweil Integrals and Weakly Henstock-Dunford Integrals

We recall a few definitions and results to make understand the chapter easily.

Definition 9.1 ([6])

(a) A function $g : [a, b] \to \mathcal{X}$ is said to be Dunford integrable on $[a, b]$ if for each $y^* \in \mathcal{X}^*$, the function y^*g is Lebesgue integrable. In this case, as a consequence of the closed graph theorem, for every measurable subset A of $[a, b]$, there exists a vector y^{**}_A in $\mathcal{X}^{**}$ such that

$$< y^*, y^{**}_A >= \int_A y^* g \text{ for all } y^* \in \mathcal{X}^*.$$

B. Hazarika, H. Kalita, *Theory of Henstock–Orlicz Spaces*, Mathematical Marvels: Texts and Monographs in the Spirit of CR Rao,
https://doi.org/10.1007/978-981-96-9548-5_9

A vector ${y_A}^{**}$ is called the Dunford integral of g on A and is denoted by $\mathcal{D} \int_A g$.

(b) A function $g : [a, b] \to \mathcal{X}$ is said to be Pettis integrable on $[a, b]$ if it is Dunford integrable on $[a, b]$ and $y_A^{**} \in \mathcal{X}$ for every measurable subset A of $[a, b]$. The Henstock integral of Banach valued functions is exactly the same way as the Dunford and Pettis are extensions of the Lebesgue integral. We define Henstock-Gel'fand integral as the style of R.A. Gordon [8] of his Denjoy-Dunford and Denjoy-Pettis integral. Also, we refer [7, 13, 18, 19, 22] for the reader related to this areas.

Definition 9.2 ([14])

(a) A function $g : [a, b] \to \mathcal{X}$ is said to be Henstock-Dunford integrable on $[a, b]$ if for each y^* in $\mathcal{X}^*$, the function y^*g is Henstock integrable on $[a, b]$ and if for every interval $\mathcal{I}$ in $[a, b]$, there exists a vector $y_{\mathcal{I}}^{**}$ in $\mathcal{X}^{**}$ such that

$$y_{\mathcal{I}}^{**}(y^*) = \int_{\mathcal{I}} y^*g \text{ for all } y^* \in \mathcal{X}^*.$$

We write $y_{\mathcal{I}_0}^{**} = HD \int_{\mathcal{I}_0} g = G(\mathcal{I}_0)$.

(b) A function $g : [a, b] \to \mathcal{X}$ is said to be Henstock-Pettis integrable on $[a, b]$ if g is Henstock-Dunford integrable on $[a, b]$ and if $y_{\mathcal{I}}^{**}$ in $\mathcal{X}$ for every interval $\mathcal{I}$ in $[a, b]$. We write

$$y_{\mathcal{I}_0}^{**} = HP \int_{\mathcal{I}_0} g = G(\mathcal{I}_0).$$

Definition 9.3 ([4, 6])

(a) A function $g : [a, b] \to \mathcal{X}^*$ is said to be Gel'fand integrable on $[a, b]$ if for each $y \in \mathcal{X}$, yg is Lebesgue integrable. In this as consequence of the closed graph theorem, for every measurable subset A of $[a, b]$ there exists y_A^* in $\mathcal{X}^*$ such that

$$< y_A^*, y >= \int_A yg \text{ for all } y^* \in \mathcal{X}^*.$$

The vector y_A^* is called the Gel'fand integral of g on A and is denoted by $\mathcal{G} \int_A g$.

(b) A function $g : [a, b] \to \mathcal{X}^*$ is said to be Henstock-Gel'fand integrable on $[a, b]$ if for each $y \in \mathcal{X}$, yg is Henstock integrable on $[a, b]$ and for every interval $\mathcal{I}$ in $[a, b]$ there exists a vector $y_{\mathcal{I}}^* \in \mathcal{X}^*$ such that $y^*(y) = \int_{\mathcal{I}} yg$.

For our work we prefer the definition of weakly Henstock-integrable like Y. Guoju, and S. Schwabik defines weakly McShane integral in [15]. We state our definitions as follows:

Definition 9.4 A function $g : [a, b] \to \mathcal{X}$ is said to be weakly Henstock integrable ($w\mathcal{H}$) on $[a, b]$ if for every $x^* \in X^*$ the real function $x^*(f)$ is Henstock integrable on $[a, b]$ and for every interval $I \subset [a, b]$, there is a $x_I \in \mathcal{X}$ such that $\int_I x^*(f) = x^*(x_I)$. We write $(wHK) \int_I f = x_I$.

Definition 9.5 A function $g : [a, b] \to \mathcal{X}$ is said to be (wHD) on $[a, b]$ if for every $x^* \in X^*$ the real function $x^*(f)$ is weak Henstock integrable on $[a, b]$ and for every interval $I \subset [a, b]$, there is a $x_I^* \in \mathcal{X}^{**}$ such that $\int_I x^*(f) = x_I^{**}(x^*), \ \forall x^* \in \mathcal{X}^*$.

Weakly McShane integral is clearly weakly Henstock integrable. Also, $g : [a, b] \to \mathcal{X}$ is scalarly integrable if for every $x^* \in \mathcal{X}^*$ the real-valued function x^*g is Lebesgue integrable. The known fact that Henstock integrable function g is measurable, and g is Lebesgue integrable if and only if g and $|g|$ are Henstock-integrable. That is, the real function x^*g of the Definition 9.4 is not Lebesgue integrable in general. This shows that weakly Henstock integrable function is not scalar integrable in general.

Definition 9.6 ([21, Definition 2.1]) A weakly measurable function $g : [a, b] \to \mathcal{X}$ is said to be determined by a weakly compact generated (WCG) if there is a weakly compact generated subspace D of $\mathcal{X}$ such that $x^*g = 0$ a.e. for all $x^* \in \mathcal{X}^*$ with $x^*|D = 0$.

Theorem 9.1 ([14]) *A function $g : [a, b] \to \mathcal{X}$ is Henstock-Dunford integrable on $[a, b]$ if and only if y^*g is Henstock integrable on $[a, b]$ for all $y^* \in \mathcal{X}^*$.*

Lemma 9.1 ([10]) *A function $g : [a, b] \to \mathbb{R}$ is Henstock integrable on $[a, b]$ equivalent to Denjoy integrable on $[a, b]$.*

Lemma 9.2 ([20]) *Suppose $\mathcal{X}$ contains no isomorphic copy of c_0 and let $g : [a, b] \to \mathcal{X}$ be $w\mathcal{H}$ scalarly integrable function. If g is weakly Henstock integrable on $[a, b]$, then it is Pettis integrable.*

Lemma 9.3 *Suppose $\mathcal{X}$ contains no isomorphic copy of c_0 and let $g : [a, b] \to \mathcal{X}$ be scalarly integrable also $w\mathcal{HD}$-integrable function on $[a, b]$, then it is Pettis integrable.*

Proof If $\mathcal{X}$ contains no copy of c_0 and $g : [a, b] \to \mathcal{X}$ is $w\mathcal{HD}$-integrable, then x^*g is weakly Henstock integral as well as scalarly integrable. Then Lemma 9.1 shows that x^*g is $w\mathcal{H}$ on $[a, b]$ also x^*g is measurable for all $x^* \in \mathcal{X}^*$. That is, g Henstock-integrable function. So g is McShane integrable and hence Pettis integrable. □

Definition 9.7 ([4, Definition 1.3], [11, Definitions 1 and 11])

(a) A function F is ACG on $\mathcal{E}$ if F is continuous on $\mathcal{E}$ and if E can be expressed as a countable union of sets on each of which F is AC.

(b) A function $g : [a, b] \to \mathcal{X}$ is Denjoy integrable on $[a, b]$ if there exists an ACG function $F : [a, b] \to \mathcal{X}$ such that $F'_{ap} = g'$ a.e. on $[a, b]$, where F'_{ap} denotes the approximate derivatives of F. In this case

$$\int_a^b g = F(b) - F(a).$$

We say g is Denjoy integrable on a subset A of $[a, b]$ if $g\chi_A$ is Denjoy integrable on $[a, b]$ and write

$$\int_A g = \int_a^b g\chi_A.$$

(c) A function $g : [a, b] \to X^*$ is said to be Denjoy-Gel'fand integrable on $[a, b]$ if for each $y \in \mathcal{X}$, yg is Denjoy integrable on $[a, b]$ and for every interval $\mathcal{I}$ in $[a, b]$ there exists a vector $y^*_{\mathcal{I}}$ called Denjoy-Gel'fand integral of g on $[a, b]$ and we denote $\mathcal{DG}\int_a^b g$.

Remark 9.1 ([12, Theorem 15.9]) A real Denjoy integrable function on $[a, b]$ is not necessarily integrable on all measurable subset of $[a, b]$. In fact, the function is absolutely integrable or equivalent to Lebesgue integral.

9.3 Results on Denjoy-Gel'fand and Henstock-Gel'fand

Proposition 9.1 *Assume $g : [a, b] \to \mathcal{X}$ is Denjoy-Gel'fand on $[a, t]$ for all $t \in [a, b)$ and for each $y \in \mathcal{X}$. If the limit $\lim_{t \to b} \int_a^t yg$ exists, then g is Denjoy-Gel'fand on $[a, b]$ and*

$$< y, \mathcal{DG}\int_a^b g >= \lim_{t \to b} < y, \mathcal{DG}\int_a^t g >$$

for each $y \in \mathcal{X}$.

Proof For the proof we follow the same technique as used in [3, Proposition 1]. □

Proposition 9.2 *Let $g : [a, b] \to \mathcal{X}$, if yg is Denjoy integrable on $[a, b]$ for each $y \in \mathcal{X}$, then each perfect set in $[a, b]$ contains a portion of which g is Gel'fand integrable.*

Proof For the proof of this proposition one can use [9, Theorem 33]. □

Proposition 9.3 *Let $g : [a, b] \to \mathcal{X}^*$ be such that yg is Denjoy integrable on $[a, b]$ for all $y \in \mathcal{X}$. Let P be a closed subset of $[a, b]$ and assume that g is Denjoy-Gel'fand integrable on each open interval J disjoint from P, then there exists a portion P_0 such that if $(\mathcal{I}_n)$ is an enumeration of the interval neighboring to P_0 then the series $\sum\limits_n \int\limits_{\mathcal{I}_n} yg$ is absolutely convergent for every $y \in \mathcal{X}$.*

Proof Using [9, Lemma 1], one can prove it. □

Proposition 9.4 *The function $g : [a, b] \to \mathcal{X}^*$ is Denjoy-Gel'fand on $[a, b]$ if and only if yg is Denjoy integrable on $[a, b]$ for all $y \in \mathcal{X}$.*

Proof The proof of this proposition follows from the same technique as used in [9, Theorem 3]. □

Theorem 9.2 *A function $g : [a, b] \to \mathcal{X}^*$ is Henstock-Gel'fand on $[a, b]$ if and only if yg is Henstock integrable on $[a, b]$ for all $y \in \mathcal{X}$.*

Proof If g is Henstock-Gel'fand integrable on $[a, b]$. Then by definition of Henstock-Gel'fand yg is Henstock integrable on $[a, b]$.
Conversely, let yg be Henstock integrable on $[a, b]$, then by Lemma 9.1, yg is Denjoy integrable on $[a, b]$ and $\mathcal{D}\int\limits_a^b yg = HK\int\limits_a^b yg$.
Then Proposition 9.4 implies that g is Denjoy-Gel'fand integrable on $[a, b]$ and for every interval I on $[a, b]$ there exists a vector $y_I^* = \mathcal{D}\int\limits_I yg$ for all $y \in \mathcal{X}$.
That gives us $y_I^* = HK\int\limits_I yg$ for all $y \in \mathcal{X}$, so g is Henstock-Gel'fand integrable on $[a, b]$. □

Theorem 9.3 *Let $\mathcal{X}$ contain no copy of c_0 and let $g : [a, b] \to \mathcal{X}$ is $w\mathcal{HD}$-integrable and scalarly integrable on $[a, b]$, then $\{y^*g : y^* \in \overline{B^*_{\mathcal{X}}}\}$ is uniformly integrable on $HK(\mu_\infty)$.*

Proof Let $g : [a, b] \to \mathcal{X}$ is $w\mathcal{HD}$-integrable on $[a, b]$. Then g is Pettis integrable. For each equi-continuous $K \subset \mathcal{X}^*$, the set $\{y^*g : y^* \in K\}$ is relatively weakly compact in $HK(\mu_\infty)$. Hence $\{y^*g : y^* \in \overline{B^*_{\mathcal{X}}}\}$ is relatively weakly compact in $HK(\mu_\infty)$. Also [5] gives $\{y^*g : y^* \in \overline{B^*_{\mathcal{X}}}\}$ is uniformly integrable in $HK(\mu_\infty)$. □

9.4 Henstock-Dunford Integrals and H-Orlicz Spaces

Theorem 9.4 *A subset A of $HK(\mu_\infty)$ is uniformly integrable if and only if there is an N-function θ with Δ' condition such that A is relatively weakly compact in $\mathcal{H}^\theta(\mu_\infty)$.*

Proof Proof is similar to [1, Theorem 2.2]. □

Theorem 9.5 *Let $(\mathcal{B}, \Sigma, \mu_\infty)$ be a finite measure space and $g : I \subseteq \mathcal{B} \to \mathcal{X}$ be Henstock-Dunford function. Then the following are equivalent:*

(i) The Henstock-Dunford integral of g is countable additive; that is the set function $HD \int g d\mu_\infty : \Sigma \to \mathcal{X}^{**}$ defined

$$\left(HD \int g d\mu_\infty\right)(\mathcal{E}) = HD \int\limits_{\mathcal{E}} g d\mu_\infty.$$

(ii) There is an N-function θ with Δ' property such that $\{y^* g : y^* \in B^*_{\mathcal{X}}\}$ is relatively weakly compact in the H-Orlicz space $\mathcal{H}^\theta(\mu_\infty)$.

Proof Let $g : [a, b] \to \mathcal{X}$ be Henstock-Dunford integrable function. Put $\gamma(\mathcal{E}) = HD \int\limits_{\mathcal{E}} g$, where $\mathcal{E} \in \Sigma$. Therefore by Theorem 9.1, we have $\gamma(E) = HK \int\limits_{\mathcal{E}} y^* g$. Let E_1, E_2 be nonoverlapping subset of $\mathcal{E}$. Then

$$\begin{aligned}
\gamma(E_1 \cup E_2) &= HD \int\limits_{E_1 \cup E_2} g \\
&= HK \int\limits_{E_1 \cup E_2} y^* g \\
&= HK \int\limits_{E_1} y^* g + HK \int\limits_{E_2} y^* g \\
&= \gamma(E_1) + \gamma(E_2)
\end{aligned}$$

and

$$\begin{aligned}
\gamma\left(\bigcup_{n=1}^{\infty} E_n\right) &= HD \int\limits_{\cup_{n=1}^{\infty} E_n} g \\
&= HK \int\limits_{\cup_{n=1}^{\infty} E_n} y^* g \\
&= \sum_{n=1}^{\infty} \gamma(E_n)
\end{aligned}$$

in norm topology of $\mathcal{X}$, for all sequence (E_n) of nonoverlapping members of field $F \subseteq [a, b]$ such that $\bigcup\limits_{n=1}^{\infty} E_n \in F$.

Now according to (see [21] for Dunford integral) γ is countable additive if and only if $T : \mathcal{X}^* \to HK(\mu_\infty)$ defined by $T(y^*) = y^*g$ is weakly compact. So, $\{y^*g : y^* \in B^*_{\mathcal{X}}\}$ is uniformly integrable in $HK(\mu_\infty)$.
Now by Theorem 9.4 it is equivalent to the existence of a N-function θ with Δ' property such that $\{y^*g : y^* \in B^*_{\mathcal{X}}\}$ is relatively compact in $\mathcal{H}^\theta(\mu_\infty)$. This completes the proof. □

Corollary 9.1 *Let $g : [a, b] \to \mathcal{X}$ be Henstock-Dunford integrable function, then the following are equivalent:*

(i) g is Henstock-Pettis integrable.
(ii) g is weakly compact generated, and there is an N-function θ with Δ' such that $\{y^*g : y^* \in B^*_{\mathcal{X}}\}$ is relatively compact in $\mathcal{H}^\theta(\mu_\infty)$.

Theorem 9.6 *A strongly measurable, then its range is essentially separable and weakly compact function $g : I \to \mathcal{X}$, and $\mathcal{X}$ contains no isomorphic copy of c_0 then is Henstock-Pettis integrable if and only if there is an N-function θ with Δ' property such that $\{y^*g : y^* \in B^*_{\mathcal{X}}\}$ is relatively weakly compact in the Orlicz space $\mathcal{H}^\theta(\mu_\infty)$.*

Proof If $g : [a, b] \to \mathcal{X}$ is strongly measurable, then its range is essentially separable and weakly compact determined [6, Theorem 2 p.42]. If g is Henstock-Pettis integrable, then each perfect set in $[a, b]$ contains a portion P which is Pettis integrable [14, Theorem 2.6]. Therefore this portion is Dunford with countable additive vector measure. Hence $\{y^*g : y^* \in B^*_{\mathcal{X}}\}$ is uniformly integrable on that portion, and consequently, there is an N-function θ with Δ' such that $\{y^*g : y^* \in B^*_{\mathcal{X}}\}$ is relatively compact in $\mathcal{L}^\theta(P)$. So,$\{y^*g : y^* \in B^*_{\mathcal{X}}\}$ is relatively compact in $\mathcal{H}^\theta(P)$.

Conversely, suppose g is strong measurable and there is an N-function θ with Δ' such that $\{y^*g : y^* \in B^*_{\mathcal{X}}\}$ is relatively compact in $\mathcal{H}^\theta(\mu_\infty)$. Since g is strongly measurable, it has range weakly compactly generated determined. As $\mathcal{H}^\theta(\mu_\infty) \subset HK(\mu_\infty)$ (see [5]). So $\{y^*g : y^* \in B^*_{\mathcal{X}}\}$ is a bounded subset of $\mathcal{H}^\theta(\mu_\infty)$. As $D_0 = \{y^*g : y^* \in B^*_{\mathcal{X}}\}$ is weakly compactly generated and g is strongly measurable, so, for each $y^* \in \mathcal{X}^*$, there exists a sequence $(\gamma)_{n=1}^\infty$ of D_0-valued simple function such that $y^*g = \lim y^*\gamma_n$ μ_∞-a.e. So, g is Pettis integrable with $\mathcal{X}$ containing no isomorphic copy of c_0. Thus g is Henstock-Pettis integrable with $\mathcal{X}$ containing no isomorphic copy of c_0. □

Corollary 9.2 *Let $g : [a, b] \to \mathcal{X}$ be Henstock-Dunford. If $p > 1$ such that $\{y^*g : y^* \in B^*_{\mathcal{X}}\}$ is bounded in $\mathcal{H}^\theta(\mu_\infty)$, then g is countably additive.*

Theorem 9.7 *If g is a strongly measurable function in $\mathcal{X}$, contains no isomorphic copy of c_0. A function $g : [a, b] \to \mathcal{X}^*$ is Henstock-Gel'fand, then there is an N-function θ with Δ' such that $\{y^*g : y^* \in B^*_{\mathcal{X}}\}$ is relatively compact in $\mathcal{H}^\theta(\mu_\infty)$.*

Proof Let $T : \mathcal{X} \to HK(\mu_\infty)$ by $T(y) = yg$. Then T is bounded linear operator.
Claim: T is weakly compact.

If $v(\mathcal{E})(y) = HG\int_{\mathcal{E}} g$, then $v(\mathcal{E})(y) = HK\int_{\mathcal{E}} yg$ (by Theorem 9.2). Then $v(\mathcal{E})$ is vector measure with countable additivity. By Pettis theorem [3, p 10] we have $\lim_{\mu_\infty(\mathcal{E})\to 0} v(\mathcal{E})(y) = 0$ for all $y \in \mathcal{X}$. That is, for $\epsilon > 0$ there exists a $\delta > 0$ such that

$$\mu_\infty(\mathcal{E}) < \delta \Rightarrow ||v(\mathcal{E})|| < \epsilon.$$

So, $\mu_\infty(\mathcal{E}) < \delta$ implies $\sup_{y\in B_{\mathcal{X}}} \int_{\mathcal{E}} |yg| d\mu_\infty < \epsilon$. By Dunford-Bartle-Hanse [3, Corollary 6, p 14], $v(\Sigma)$ is relatively weakly compact. Therefore $\left\{\int_{\mathcal{E}} yg : \mathcal{E} \in \Sigma, y \in B_{\mathcal{X}}\right\}$ is bounded in some field F. So, $\sup_{y\in B_{\mathcal{X}}} \int_{\Omega} |yg| d\mu_\infty < \infty$.

Thus $\{yg : y \in B_{\mathcal{X}}\}$ is uniformly integrable. By de la Vallée Poussin theorem $\{yg : y \in B_{\mathcal{X}}\}$ is relatively weakly compact in $\mathcal{H}^\theta(\mu_\infty)$ with Δ' condition. □

Theorem 9.8 *For a real Banach space $\mathcal{X}$ contains no isomorphic copy of c_0, if $g : [a, b] \to \mathcal{X}$ is $w\mathcal{H}$-integrable function on $[a, b]$, then $\{y^*g : y^* \in \overline{B^*_{\mathcal{X}}}\}$ is relatively weakly compact in certain separable Orlicz space $\mathcal{H}^\theta(\mu_\infty)$.*

Proof For a real Banach space $\mathcal{X}$ contains no isomorphic copy of c_0, if $g : [a, b] \to \mathcal{X}$ is $w\mathcal{H}$-integrable function on $[a, b]$, then $\{y^*g : y^* \in \overline{B^*_{\mathcal{X}}}\}$ is uniformly integrable in $HK(\mu_\infty)$. Theorem 9.4 gives $\{y^*g : y^* \in \overline{B^*_{\mathcal{X}}}\}$ is relatively weakly compact in certain separable Orlicz space $\mathcal{H}^\theta(\mu_\infty)$. As $\mathcal{X}$ is reflective, it is true for $\{y^*g : y^* \in B^*_{\mathcal{X}}\}$. □

Remark 9.2 From Lemma 9.1 countable additivity of $w\mathcal{H}$-integrable function can be determined easily.

References

1. Alexopoulos, J.: De la Vallée Poussin's theorem and weakly compact sets in Orlicz spaces. Quaestiones Mathematicae **17**(2), 231–248 (1994)
2. Alexiewicz, A.: Linear functionals on Denjoy-integrable functions. Collectanea Mathematica **1**, 289–293 (1948)
3. Barcenas, D., Finol, C.E.: On vector measures, Uniformly integrable and Orlicz Spaces in vector measures. Integration and related Topics: Operator Theory: Advances and Applications, vol. 201, pp. 51–57. Birkhaüser Verlag Basel (2010)
4. Bongiorno, B., Di Piazza, L., Musial, K.: Differentiation of additive interval Measure with values in a conjugate Banach Space. Functiones et Approximatio **50**(1), 169–180 (2014)
5. del Campo, R., Fernández, A., Mayoral, F., Naranjo, F.: The de la Vallée-Poussin theorem and Orlicz spaces associated to vector measure. J. Math. Anal. Appl. **470**, 279–291 (2019)
6. Diestel, J., Uhl, J.J.: Vector measures. American Mathematical Society, Providence (1977)
7. Dunford, N., Schwartz, J.T.: Linear Operators Part I. Wiley-Interscience, New York (1988)
8. Fong, C.K.: A continuous version of Orlicz-Pettis theorem via vector-valued Henstock-Kurzweil integrals. Can. Math. Bull. **24**(2), 169–176 (1981)

9. Gamez, J.L., Mendoza, J.: Denjoy-Dunford and Denjoy-Pettis integral. Studia Mathematica **130**(2), 115–133 (1998)
10. Gill, T.L., Zachary, W.W.: Functional Analysis and Feynman Operator Calculus. Springer, Cham (2016)
11. Gordon, R.A.: The Denjoy extension of the Bochner, Pettis, and Dunford integrals. Studia Mathematica **92**(1), 73–91 (1989)
12. Gordon, R.: The Integrals of Lebesgue, Denjoy, Perron, and Henstock. American Mathematical Society, Providence (1994)
13. Guoju, Y.: On the Henstrock-Kurzweil-Dunford and Kurzweil-Henstock-Pettis. Rocky Mount. J. Math. **39**(4), 1233–1244 (2009)
14. Guoju, Y., Tianguing, A.N.: On Henstock-Dunford and Henstock-Pettis integral. Int. J. Math. Math. Sci. **25**(7), 467–478 (2001)
15. Guoju, Y., Schwabik, S.: The Mcshane and weak Mcshane integrals of Banach space-valued functions defined on $\mathbb{R}^m$. Math. Notes Miskolc **2**(2), 127–136 (2001)
16. Hazarika, B., Kalita, H.: Henstock-Orlicz spaces and its dense spaces. Asian Eur. J. Math. **14**(7), 1–17 (2021)
17. Lindenstrauss, J., Tzafriri, L.: Classical Banach Spaces II, Function Spaces. A Series of Modern Surveys in Mathematics. Springer, Berlin-Heidelberg-New York (1979)
18. Morrison, T.J.: A note on Denjoy integrability of Abstractly valued functions. Proc. Am. Math. Soc. **61**(2), 385–386 (1976)
19. Musiał, K.: Pettis Integral. In: Pap, E. (ed.) Handbook of Measure Theory, pp. 531–568. North Holland, Amsterdam (2002)
20. Saadoune, M., Sayyad, R.: From weak Henstock to weak McShane integrability. Real Anal. Exchange **38**(2), 447–468 (2012–2013)
21. Stefansson, A.G.: Pettis integrability. Trans. Am. Math. Soc. **330**(1), 401–418 (1992)
22. Stefansson, G.F.: L^1 of a vector measure. Le Matematiche **48**, 219–234 (1993)

Chapter 10
Modular Convergence in H-Orlicz Spaces of Banach Valued Functions

10.1 Introduction

Let $(\mathcal{B}, \Sigma, \mu_\infty)$ be a measure space, where $\mathcal{B}$ is an abstract set, Σ is a σ-algebra of subsets of the set $\mathcal{B}$, μ_∞ is a σ-finite, positive, complete measure on Σ. $\mathcal{X}$ is a Banach space. Let $M_\mathcal{X}$ be a set of all μ_∞-measurable functions $\xi : \xi^i(t)$ $(t \in \mathcal{B},\ i = 1, 2, ..)$, where $f^i(t)$ are real-valued function defined on $\mathcal{B}$ with values in $\mathcal{X}$.

Definition 10.1 ([6, Definition 2.1]) A function $\theta : \mathcal{X} \times \mathcal{B} \to [0, \infty]$ is called an N-function if it satisfies the following conditions. There exists a set $\mathcal{B}_1 \in \Sigma,\ \mu_\infty(\Sigma \setminus \mathcal{B}_1) = 0$, such that

(a) $\theta(., .)$ is $\mathfrak{B} \times \Sigma$ measurable, where $\mathfrak{B}$ denotes the σ-algebra of Borel subsets of $\mathcal{X}$.
(b) $\theta(., t)$ is lower semi-continuous on X for every $t \in \mathcal{B}_1$.
(c) $\theta(., t)$ is convex for every $t \in \mathcal{B}_1$.
(d) $\theta(0, t) = 0$ and $\theta(\xi(t), t) = \theta(-\xi(t), t)$ for every $\xi \in M_\mathcal{X},\ t \in \mathcal{B}_1$.
(e) there exist μ_∞-measurable functions $\alpha : \mathcal{B}_1 \to (0, \infty)$ and $\lambda : \mathcal{B}_1 \to (0, \infty)$ such that $||\xi(t)|| \geq \lambda(t)$ implies $\theta(\xi(t), t) \geq \alpha(t)$ for every $t \in \mathcal{B}_1$.
(f) there exist μ_∞-measurable functions $\rho : \mathcal{B}_1 \to (0, \infty)$ and $\rho_0 : \mathcal{B}_1 \to (0, \infty)$ such that $||\xi(t)|| \leq \rho(t)$ implies $\theta(\xi(t), t) \leq \rho_0(t)$ for every $t \in \mathcal{B}_1$.

If θ fulfills conditions $(a) - (d)$, then θ is called N''-function, if it satisfies condition $(a) - (e)$, it is called N'-function. Recall that an N'-function θ is called an N-function if $\mathcal{X}$ is separable, reflexive Banach space if for almost $t \in \mathcal{B}$,

$$\theta(x, t) \leq \xi(t) < \infty \ for \ ||x|| \leq \rho(t),\ \xi(t),\ \rho(t) \in (0, \infty) \ holds.$$

A function $\xi : \mathcal{B} \to \mathcal{X}$ is M-measurable if there is a sequence of simple functions from $S_\mathcal{X}(\Sigma)$ converges to ξ (M-a.e.), where $S_\mathcal{X}(\Sigma)$ is a set of all $\mathcal{X}$-valued simple

B. Hazarika, H. Kalita, *Theory of Henstock–Orlicz Spaces*, Mathematical Marvels: Texts and Monographs in the Spirit of CR Rao,
https://doi.org/10.1007/978-981-96-9548-5_10

functions. M denotes a σ-bounded family of positive measures defined on Σ. This means for each $E \in \Sigma$, there exists a pairwise disjoint collection $\{E_i\}_{i=1}^{\infty}$, $E_i \in \Sigma$ such that $E = \bigcup_{i=1}^{\infty} E_i$ and $S_{\mathcal{X}}(E)$ is a set of all $\mathcal{X}$-valued simple functions.

Definition 10.2 [9] Given a linear space $\mathcal{X}$, a functional $\rho(.)$ defined on $\mathcal{X}$ with values $-\infty < \rho(x) \leq \infty$ is called a modular if the following conditions hold:

(a) $\rho(x) = 0$ if and only if $x = 0$
(b) $\rho(-x) = \rho(x)$
(c) $\rho(\alpha x + \beta y) \leq \rho(x) + \rho(y)$ for every $\alpha, \beta \geq 0,\ \alpha + \beta = 1$

Definition 10.3 [9] A sequence $(x_n) \subset \mathcal{X}$ is said to be modular convergent to $x \in \mathcal{X}$ if there exists a number $\alpha > 0$ (depending on the sequence (x_n)) such that $\rho(\alpha(x_n - x)) \to 0$ as $n \to \infty$.

Definition 10.4 ([10, Definition 2.2]) Let $\theta(t, \xi(t))$ be a generalized Young function and let $L^{\theta}(\mathcal{X}, \mu_{\infty})$ be its associated Orlicz class. The closure $\mathcal{L}^{\theta}(\mathcal{X}, \mu_{\infty})$ of $\mathcal{L}^{\theta}(\mathcal{X}, \mu_{\infty})$ under positive scalar multiplication is a vector-valued Orlicz space.

The space $\mathcal{L}^{\theta}(\mathcal{X}, \mu_{\infty})$ is a Banach space with Luxemburg type norm

$$||\xi||_{\mathcal{L}^{\theta}} = \inf\left\{\alpha > 0 : L\int_{\mathcal{B}} \theta\left(t, \frac{\xi(t)}{\alpha}\right) d\mu_{\infty} \leq 1\right\}.$$

Recalling convergence of a sequence of function ξ_n in the normed vector space $\mathcal{L}^{\theta}(\mathcal{X}, \mu_{\infty})$ to ξ in the following way:

$$\xi_n \to \xi \text{ if } ||\xi_n - \xi||_{\mathcal{L}^{\theta}} \to 0 \text{ as } n \to \infty.$$

10.2 Henstock-Kurzweil Integrable Function Spaces $HK(\mathcal{X}, \mu_{\infty})$

In this section of the monograph we discuss about the Henstock-Kurzweil integrable function space. We have left off the basic results of the space. We discuss only those results that are closely connected for the next section.

A finite collection $D = \{D_i : i = 1, 2, \ldots, n\}$ of mutually disjoint elements of Σ is called sub-partition of $\mathcal{B}$ if $\mathcal{B} = \bigcup_{i=1}^{n} D_i$. In this case $D = \{D_i : i = 1, 2, \ldots, n\}$ is called partitions of $\mathcal{B}$.

Let $\widehat{D} = \{(D_i, d_i) : i = 1, 2, \ldots, n\}$ and $d_i \in D_i$. Clearly the norm of the sub-partition $D = \{D_i\}_{i=1}^{n}$ is $||D|| = \sup\{\mu_{\infty}(D_i) : i = 1, 2, \ldots, n\}$. We assume the collection of all tagged sub-partitions of $\mathcal{B}$ will be $\mathfrak{Y}$. If $\widehat{D_1},\ \widehat{D_2} \in \mathfrak{Y}$, then the sub-partition

$$D_1 \vee D_2 = (D \setminus \cup D_2) \cup \{A \cap B : A \in D_1,\ B \in D_2\}.$$

Clearly $(\mathfrak{Y}, \gg)$ is a directed set. Let $\xi : \mathcal{B} \to \mathcal{X}$ and $\widehat{D} = \{(\widehat{D_1}, d_i) : i = 1, 2, .., n\} \in \mathfrak{Y}$, then we construct the Riemann sum of ξ corresponding to $\widehat{D}$ as

$$S(\xi, \widehat{D}) = \sum_{i=1}^{n} \xi(d_i)\mu_\infty(D_i).$$

As $(\mathfrak{Y}, \gg)$ is directed set, the function $\widehat{D} = \{(D_i, d_i) : i = 1, 2, .., n\} \to S(\xi, \widehat{D})$ defines a net $S : \mathfrak{Y} \to \mathcal{X}$.

Definition 10.5 ([5]) A function $\xi : \mathcal{P} \subset \Sigma \to \mathcal{X}$ is said to be Henstock-Kurzweil integrable on the set $\mathcal{P} \in \Sigma$ if there is an element $I_{\mathcal{P}} = \int_{\mathcal{P}} f d\mu_\infty$ such that for every $\epsilon > 0$ there exist $\widehat{D_0} \in \mathfrak{Y}$ and $\widehat{D} \gg \widehat{D_0}$ in $\mathfrak{Y}$, then

$$||S(\xi, \widehat{D}) - \int_\Sigma \xi d\mu_\infty|| \leq \epsilon.$$

We call net limits of the integral $\xi : \Sigma \to \mathcal{X}$ as

$$\int_\Sigma \xi d\mu_\infty = \lim \left(S(\xi, \widehat{D})_{D\in(\mathfrak{Y},\gg)}\right).$$

These integrals are unique, linear also for all $\mathcal{X}$-valued Henstock integrable. Let $HK(\Sigma, \mu_\infty, \mathcal{X})$ or in brief $HK(\mathcal{X}, \mu_\infty)$ the set of all $\mathcal{X}$-valued μ_∞-Henstock-Kurzweil integrable functions.

Definition 10.6 For each $\xi : \Sigma \to \mathcal{X}$

$$||\xi||_{\mathfrak{Y}} = \sup\{||S(\xi, \widehat{D})|| : \widehat{D} \in \mathfrak{Y}\}. \tag{10.1}$$

Theorem 10.1 *The expression (10.1) is a norm on $HK(\mathcal{X}, \mu_\infty)$.*

Proof For any vector-valued function $\xi_1, \xi_2 \in HK(X, \mu_\infty)$, we have the nonnegativity as follows:

(i)

$$\begin{aligned}||\xi_1||_{\mathfrak{Y}} &= \sup\left\{||S(\xi_1, \widehat{D})|| \ : \ \widehat{D} \in \mathfrak{Y}\right\}\\ &= \sup\left\{||\sum_{i=1}^{n} \xi_1(d_i)\mu_\infty(D_i)|| \ : \ \widehat{D} = (D_i, d_i),\ i = 1, 2, ..\right\}\\ &\geq 0.\end{aligned}$$

Therefore, $||\xi_1||_{\mathfrak{Y}} \geq 0$.

(ii) Now, $||\xi_1||_{\mathfrak{Y}} = 0$ if and only if

$$
\begin{aligned}
||\xi_1||_{\mathfrak{Y}} &= \sup\left\{||S(\xi_1, \widehat{D})|| : \widehat{D} \in \mathfrak{Y}\right\} \\
&= \sup\left\{||\sum_{i=1}^{n} \xi_1(d_i)\mu_\infty(D_i)|| : \widehat{D} = (D_i, d_i),\ i = 1, 2, \ldots\right\} \\
&= \sum_{i=1}^{n} \xi_1(d_i)\mu_\infty(D_i) \\
&= 0.
\end{aligned}
$$

Therefore, $||\xi_1||_{\mathfrak{Y}} = 0$ if and only if $\xi_1 = 0$.

(iii) For a scalar α, we have

$$
\begin{aligned}
||\alpha\xi_1||_{\mathfrak{Y}} &= \sup\left\{||S(\alpha\xi_1, \widehat{D})|| \ : \ \widehat{D} \in \mathfrak{Y}\right\} \\
&= \sup\left\{||\alpha S(\xi_1, \widehat{D})|| \ : \ \widehat{D} \in \mathfrak{Y}\right\} \text{ [2, page 199]} \\
&= |\alpha| \sup\left\{||S(\xi_1, \widehat{D})|| : \widehat{D} \in \mathfrak{Y}\right\} \\
&= |\alpha| ||\xi_1||_{\mathfrak{Y}}.
\end{aligned}
$$

(iv) For triangle inequality,

$$
\begin{aligned}
||\xi_1 + \xi_2||_{\mathfrak{Y}} &= \sup\left\{||S(\xi_1 + \xi_2, \widehat{D})|| \ : \ \widehat{D} \in \mathfrak{Y}\right\} \\
&= \sup\left\{||S(\xi_1, \widehat{D}) + S(\xi_2, \widehat{D})|| \ : \ \widehat{D} \in \mathfrak{Y}\right\} \text{ [2, page 199]} \\
&\leq \sup\left\{||S(\xi_1, \widehat{D})|| \ : \ \widehat{D} \in \mathfrak{Y}\right\} + \sup\left\{||S(\xi_2, \widehat{D})|| \ : \ \widehat{D} \in \mathfrak{Y}\right\} \\
&\leq ||\xi_1||_{\mathfrak{Y}} + ||\xi_2||_{\mathfrak{Y}}.
\end{aligned}
$$

□

Then clearly $||\xi||_{\mathfrak{Y}} < \infty$, and the norm (10.1) is equivalent to the Alexiewicz norm for the Henstock integrable function spaces.

Theorem 10.2 *Let* $\xi : \Sigma \to X$. *If* $\xi = 0$ μ_∞*-a.e. on* Σ, *then* $\xi \in HK(\mathcal{X}, \mu_\infty)$ *and* $\int_\Sigma \xi d\mu_\infty = 0$.

Proof Let $\mathcal{B} = \{z \in \Sigma : \xi(z) \neq 0\}$. For each positive n, assume

$$\mathcal{B}_n = \{z \in \mathcal{B} : n - 1 \leq ||\xi(z)|| < n\}.$$

Choosing an open set O_n for each n such that $\mathcal{B}_n \subseteq O_n$ and $\mu_\infty(O_n) < \frac{\epsilon}{n.2^n}$. For a positive function

$$\xi(x) = \begin{cases} 1\, if\ x \in \Sigma - \mathcal{B} \\ e\, if\ x \in \mathcal{B}_n, \end{cases}$$

where $e(z, \mathcal{B}_n) = \inf\{|y - z| : y \in \mathcal{B}_n\}$, then $e(z, \mathcal{B}_n) > 0$ if $x \notin \mathcal{B}_n$ and $\mathcal{B}_n$ is closed. If D_n is the subset of tagged partition D of Σ that is subordinate to ξ in $\mathcal{B}_n$.

Now,

$$\begin{aligned} ||S(\xi, \widehat{D})|| &\leq \sum_{n=1}^{\infty} ||S(\xi, \widehat{D})|| \\ &< \sum_{n=1}^{\infty} n\nu(O_n) \\ &< \sum_{n=1}^{\infty} \epsilon.2^{-n} \\ &= \epsilon. \end{aligned}$$

So, $\xi \in HK(\mathcal{X}, \mu_\infty)$ and $\int_\Sigma \xi = 0$. □

Corollary 10.1 *Let $\xi, g \in HK(\mathcal{X}, \mu_\infty)$ be μ_∞-essentially equal, then $\int_\Sigma \xi d\mu_\infty = \int_\Sigma g d\mu_\infty$.*

Theorem 10.3 *Let $(\Sigma, \mathcal{B}, \mu_\infty)$ be a finite measure, $\mathcal{X}$ is a Banach space, then $\big(HK(\mathcal{X}, \mu_\infty), ||.||_{\mathfrak{Y}}\big)$ is complete.*

Proof Let us assume (ξ_n) be a Cauchy sequence in $HK(\mathcal{X}, \mu_\infty)$, then for each $\epsilon > 0$ we can find a natural number N such that $m, n > N$ such that

$$\sup_{\widehat{D} \in \mathfrak{Y}} ||S(\xi_m - \xi_n, \widehat{D})|| \leq \epsilon.$$

If $\overline{w} \in \Sigma$, under the assumption of sub-partition $\widehat{D} = \{(\Sigma, \overline{w})\}$, we get $||\xi_n(\overline{w}) - \xi_m(\overline{w})|| \leq \mu_\infty\big(\Sigma\big)\epsilon$. From our hypothesis $\epsilon > 0$, $\mu_\infty(\Sigma) < \infty$, we find $\big(\xi_n(\overline{w})\big)$ is Cauchy sequence of $\mathcal{X}$. Again, since $\mathcal{X}$ is a Banach space, it is easy to define $\overline{w} \to \xi(\overline{w}) = \lim_{n\to\infty} \xi_n(\overline{w})$.

Using the concept of the Riemann sum over sub-partitions $\widehat{D}_1$ and $\widehat{D}_2$ in $\mathfrak{Y}$ with $\widehat{D} = \widehat{D}_1 \cup \widehat{D}_2$ we get the following:

$$\begin{aligned}
&||\int_\Sigma \xi_n d\mu_\infty - \int_\Sigma \xi_m d\mu_\infty|| \\
&\leq ||\int_\Sigma \xi_n d\mu_\infty - S(\xi_n, \widehat{D})|| + ||S(\xi_n, \widehat{D}) - S(\xi_m, \widehat{D})|| \\
&+ ||\int_\Sigma \xi_m d\mu_\infty - S(\xi_n, \widehat{D})|| \\
&< 3\epsilon.
\end{aligned}$$

This gives, for $N \in \mathbb{N}$, $||\int_\Sigma \xi_n d\mu_\infty - \int_\Sigma \xi_m d\mu_\infty|| < \epsilon$ for $m, n \geq N$. So, we can conclude $\int_\Sigma \xi_n d\mu_\infty$ is Cauchy sequence in $\mathcal{X}$. Say $\int_\Sigma \xi_n d\mu_\infty$ converges to $x \in X$.

Again,

$$\begin{aligned}
&||S(\xi, \widehat{D}) - x|| \leq ||S(\xi, \widehat{D}) - S(\xi_n, \widehat{D})|| + ||S(\xi_n, \widehat{D}) - S(\xi_m, \widehat{D})|| \\
&+ ||S(\xi_m, \widehat{D}) - \int_\Sigma \xi_m d\mu_\infty|| + ||\int_\Sigma \xi_m d\mu_\infty - x|| \\
&< 4\epsilon.
\end{aligned}$$

So, $\xi \in HK(\mathcal{X}, \mu_\infty)$ and $x = \int_\Sigma \xi d\mu_\infty$. □

Theorem 10.4 *If $\mathcal{X}$ is weakly sequentially complete, then $HK(\mathcal{X}, \mu_\infty)$ is also sequentially complete.*

Proof The proof is similar to [3, Theorem 2.2]. □

Theorem 10.5 *Let $\xi : \mathbb{J} \subseteq \mathcal{B} \rightarrow \mathcal{X}$ be in $HK(\mathcal{X}, \mu_\infty)$, then $\theta(t, \xi(t)) \in HK(\mathcal{X}, \mu_\infty)$ for all $t \in \mathbb{J}$.*

Proof Let $\xi : \mathbb{J} \subseteq \mathcal{B} \rightarrow \mathcal{X}$ be measurable function, then for all $\epsilon > 0$ there exists a $\delta : \mathbb{J} \rightarrow (0, +\infty)$ such that

$$\left\| \sum_{(\mathbb{J},t)\in\pi} \sum_{(\mathbb{J}',t')\in\pi'} [\xi(t) - \xi(t')]\mu_\infty(\mathbb{J} \cap \mathbb{I}') \right\|_X < \epsilon$$

for all partitions π and π' of $\mathbb{J}$ finer than δ.

As θ is Young function. So, $\theta(t, \xi(t)) \rightarrow \infty$ as $t \rightarrow \infty$. Our claim is $\theta(t, \xi(t))$ is Henstock-Kurzweil integrable. Since Young function by definition is an extended real Borel function. So, $\theta(t, \xi(t))$ is measurable. If π and $\pi^{'}$ are both partitions of the same interval of $\mathbb{J}$, then for any subinterval $\mathbb{J}_0$ of $\mathbb{J}_1$ we can write

$$\mu_{\infty}(\mathbb{J}) = \sum_{(\mathbb{J}^{'}, t^{'}) \in \pi^{'}} \mu_{\infty}(\mathbb{J} \cap \mathbb{J}^{'}).$$

So,

$$\sum_{(\mathbb{J}, t) \in \pi} \theta(t, \xi(t))(t)\mu_{\infty}(\mathbb{J}) = \sum_{(\mathbb{J}, t) \in \pi} \sum_{(\mathbb{J}^{'}, t^{'}) \in \pi^{'}} \theta(t, \xi(t))\overline{\mu_{\infty}}(\mathbb{J} \cap \mathbb{J}^{'})$$

$$\text{i.e.,} \left\| \sum_{(\mathbb{J}, t) \in \pi} \theta(t, \xi(t))\overline{\mu_{\infty}}(\mathbb{J}) - \sum_{(\mathbb{J}^{'}, t^{'}) \in \pi^{'}} \theta(t, \xi(t))\overline{\mu_{\infty}}(\mathbb{J}) \right\|_{\mathcal{X}} < \epsilon.$$

Thus, $\theta(t, \xi(t))$ is in $HK(\mathcal{X}, \mu_{\infty})$. □

Corollary 10.2 *For all functions $\xi \in M_{\mathcal{X}}$ there exists a constant $k > 0$ such that*

$$\rho(k\xi) = \int_{\mathcal{B}} \theta(t, k\xi(t))d\mu_{\infty} \in HK(\mathcal{X}, \mu_{\infty}).$$

10.3 H-Orlicz Spaces $\mathcal{H}^{\theta}(\mathcal{X}, \mu_{\infty})$

In this section, we initiate the study of H-Orlicz spaces associated with the modular of functions. Let $M_{\mathcal{X}}$ be the set of all real-valued (or complex-valued), Σ-measurable and finite μ_{∞}-a.e. functions on $\mathcal{B}$, with equality μ_{∞}-a.e. Clearly $\theta(t, \xi(t))$ is Σ-measurable function of $t \in \mathcal{B}$ for every $f \in M_{\mathcal{X}}$ we define

$$\rho(\xi) = HK \int_{\mathcal{B}} \theta(t, \xi(t))d\mu_{\infty}. \tag{10.2}$$

Clearly, $\rho(x)$ of Eq. (10.2) is a modular in $M_{\mathcal{X}}$.

We define the H-Orlicz class as follows. For some $k > 0$

$$H^{\theta}(\mathcal{X}, \mu_{\infty}) = \left\{ \xi : \mathcal{B} \to \mathcal{X} \text{ measurable} : \int_{\mathcal{B}} \theta\big(t, k\xi(t)\big)d\mu_{\infty} \in H(\mathcal{X}, \mu_{\infty}) \right\}.$$

It is very straightforward that:

$$H^\theta(\mathcal{X}, \mu_\infty)$$
$$= \left\{ \xi : \mathcal{B} \to \mathcal{X} \text{ measurable} : \int_{\mathcal{B}} \theta\big(t, k\xi(t)\big) d\mu_\infty \in H(\mathcal{X}, \mu_\infty) \right\} \to 0$$

as $k \to 0+$.

Theorem 10.6 *The H-Orlicz class $H^\theta(\mathcal{X}, \mu_\infty)$ is a convex set of functions. That is, for given $\alpha, \beta \geq 0,\ \alpha + \beta = 1,\ \rho(\alpha\xi + \beta g) \leq \alpha\rho(\xi) + \beta\rho(g)$.*

Proof The proof is similar to [4, Theorem 2.1]. □

Theorem 10.7 *The H-Orlicz class $H^\theta(\mathcal{X}, \mu_\infty)$ is linear if and only if $H^\theta(\mathcal{X}, \mu_\infty)$ is closed under positive scalar multiplication.*

Proof Let $H^\theta(\mathcal{X}, \mu_\infty)$ be a linear space, then clearly it is closed under positive scalar multiplication.

Conversely, assume if $H^\theta(\mathcal{X}, \mu_\infty)$ is closed under positive scalar multiplication. We will prove $H^\theta(\mathcal{X}, \mu_\infty)$ is linear space. From the definition of $\theta(t, \xi(t))$, we find $\theta(t, \xi(t)) = \theta(t, -\xi(t))$, this means $-\xi \in H^\theta(\mathcal{X}, \mu_\infty)$. Assume $\alpha,\ \beta > 0$ are real numbers and $\xi, g \in H^\theta(\mathcal{X}, \mu_\infty)$.

Case I: Let $\alpha,\ \beta > 0$, then for each $t \in \mathcal{B}$ and convexity of $\theta\big(t, \xi(t)\big)$ we have

$$\theta\left(t, \frac{\alpha\xi(t) + \beta g(t)}{\alpha + \beta}\right) = \theta\left(t, \frac{|\alpha|\xi(t)}{|\alpha| + |\beta|}\right) + \theta\left(t, \frac{|\beta| g(t)}{|\alpha| + |\beta|}\right). \tag{10.3}$$

By the assumption $\alpha + \beta > 0$. The right sides of Eq. (10.3) are in $HK(\mathcal{X}, \mu_\infty)$, so we can conclude $\alpha\xi + \beta g \in H^\theta(\mathcal{X}, \mu_\infty)$.

Case II: If $\alpha\beta < 0$ with $\alpha < 0 < \beta$, then

$$\theta\left(t, \frac{|\alpha|(-\xi(t)) + \beta g(t)}{|\alpha| + \beta}\right) = \theta\left(t, \frac{|\alpha|(-\xi(t))}{|\alpha| + \beta}\right) + \theta\left(t, \frac{\beta g(t)}{|\alpha| + \beta}\right). \tag{10.4}$$

By the assumption $|\alpha| + \beta > 0$. The right sides of Eq. (10.4) are in $HK(\mathcal{X}, \mu_\infty)$, so we have $\alpha\xi + \beta g \in H^\theta(\mathcal{X}, \mu_\infty)$. □

We define vector-valued H-Orlicz space as below.

Definition 10.7 The closure of H-Orlicz class $\mathcal{H}^\theta(\mathcal{X}, \mu_\infty)$ under positive scalar multiplication of a vector-valued generalized Young function $\theta(t, \xi(t))$ is called vector-valued H-Orlicz space, denoted as $\mathcal{H}^\theta(\mathcal{X}, \mu_\infty)$. That is,

$$\mathcal{H}^\theta(\mathcal{X}, \mu_\infty) = \big\{\xi \in H^\theta(\mathcal{X}, \mu_\infty) :\ c\xi \in H^\theta(\mathcal{X}, \mu_\infty)\big\}.$$

We define the norm of $\mathcal{H}^\theta(\mathcal{X}, \mu_\infty)$ as follows:

$$||\xi||_{(\mathcal{X},\mu_\infty)} = \inf\left\{\alpha > 0 : \ HK \int_{\mathcal{B}} \theta\left(t, \frac{\xi(t)}{\alpha}\right) d\mu_\infty \leq 1\right\}. \tag{10.5}$$

It is very clear that $\left(\mathcal{H}^\theta(\mathcal{X}, \mu_\infty), ||.||_{(\mathcal{X},\mu_\infty)}\right)$ is a Banach space with norm (10.5).

Theorem 10.8 *The classical Orlicz space $\mathcal{L}^\theta(\mathcal{X}, \mu_\infty)$ is a dense subspace of $\mathcal{H}^\theta(\mathcal{X}, \mu_\infty)$ as continuous dense embeddings. That is, $\mathcal{L}^\theta(\mathcal{X}, \mu_\infty) \hookrightarrow \mathcal{H}^\theta(\mathcal{X}, \mu_\infty)$ is continuous dense embeddings.*

Proof Let $h \in \mathcal{L}^\theta(\mathcal{X}, \mu_\infty)$. Then $h \in \mathcal{L}^1(\mathcal{X}, \mu_\infty)$ with $||f||_{\mathcal{L}^\theta} < \infty$. Then for some $\alpha > 0$, we have

$$\begin{aligned}
&\inf\left\{\alpha > 0 : HK \int_{\mathcal{B}} \theta\left(t, \frac{h(t)}{\alpha}\right) d\mu_\infty\right\} \\
&\leq \inf\left\{\alpha > 0 : L \int_{\mathcal{B}} \theta\left(t, \frac{h(t)}{\alpha}\right) d\mu_\infty\right\} \\
&\leq 1.
\end{aligned}$$

So, for some $\alpha > 0$, $\inf\left\{\alpha > 0 : L \int_{\mathcal{B}} \theta\left(t, \frac{h(t)}{k}\right) d\mu_\infty \leq 1\right\}$, we get the following

$$\inf\left\{\alpha > 0 : HK \int_{\mathcal{B}} \theta\left(t, \frac{h(t)}{\alpha}\right) d\mu_\infty \leq 1\right\}.$$

Hence $h \in \mathcal{H}^\theta(\mathcal{X}, \mu_\infty)$ with $||h||_{(\mathcal{X},\mu_\infty)} \leq ||h||_{\mathcal{L}}$. Hence the proof. □

Theorem 10.9 *Suppose $\mu_\infty(\mathcal{X}) < \infty$ and μ_∞ is bounded, then $\mathcal{H}^\theta(\mathcal{X}, \mu_\infty) \hookrightarrow \mathcal{L}^1(\mathcal{X}, \mu_\infty)$ is continuous.*

Corollary 10.3 *Suppose $\mu_\infty(\mathcal{X}) < \infty$ and μ_∞ is bounded, then $\mathcal{H}^\theta(\mathcal{X}, \mu_\infty) \hookrightarrow HK(\mathcal{X}, \mu_\infty)$ is continuous.*

10.4 Modular and Norm Convergence of $\mathcal{H}^\theta(\mathcal{X}, \mu_\infty)$

It is well known that norm convergence implies modular convergence in classical Orlicz spaces (see [8, Page 9]). In this section, we discuss the relationship of modular and norm convergent of $\mathcal{H}^\theta(\mathcal{X}, \mu_\infty)$ and $\mathcal{L}^\theta(\mathcal{X}, \mu_\infty)$

Definition 10.8 We say that a sequence $(\xi_n) \in \mathcal{H}^\theta(\mathcal{X}, \mu_\infty)$ is modular convergent to $\xi \in \mathcal{H}^\theta(\mathcal{X}, \mu_\infty)$ if there exists a constant $c > 0$ such that $\rho\big(c(\xi_n - \xi)\big) \to 0$ as $n \to \infty$.

Theorem 10.10 *Modular convergent in $\mathcal{L}^{\theta}(\mathcal{X}, \mu_{\infty})$ implies modular convergent in $\mathcal{H}^{\theta}(\mathcal{X}, \mu_{\infty})$.*

Proof Let $\xi_n(.) \in \mathcal{L}^{\theta}(\mathcal{X}, \mu_{\infty})$ be modular convergent to $\xi(.) \in \mathcal{L}^{\theta}(\mathcal{X}, \mu_{\infty})$. Then there exists a constant $c > 0$ such that $\lim_{n\to\infty} \rho[c(\xi_n - \xi)] = 0$. Since $\mathcal{L}^{\theta}(\mathcal{X}, \mu_{\infty})$ is subset of $\mathcal{H}^{\theta}(\mathcal{X}, \mu_{\infty})$ as continuous dense embedding, so $\xi_n(.) \in H^{\theta}(\mathcal{X}, \mu_{\infty})$. Using definition 10.8, $\xi_n(.)$ is modular convergent to $\xi(.) \in \mathcal{H}^{\theta}(\mathcal{X}, \mu_{\infty})$. □

Remarkk 10.1 The known fact norm convergence is modular convergence in $\mathcal{L}^{\theta}(\mathcal{X}, \mu_{\infty})$ along with (Theorem 10.10), norm convergence in $\mathcal{L}^{\theta}(\mathcal{X}, \mu_{\infty})$ is modular convergence in $\mathcal{H}^{\theta}(\mathcal{X}, \mu_{\infty})$.

Now we will check the relationship of norm convergence

Theorem 10.11 *Norm convergent in $\mathcal{L}^{\theta}(\mathcal{X}, \mu_{\infty})$ are norm convergent in $\mathcal{H}^{\theta}(\mathcal{X}, \mu_{\infty})$.*

Proof Let $\xi_n \in \mathcal{L}^{\theta}(\mathcal{X}, \mu_{\infty})$ such that $\xi_n \to f$ in $\mathcal{L}^{\theta}(\mathcal{X}, \mu_{\infty})$ in the way that $||\xi_n - \xi||_{\mathcal{L}^{\theta}} \to 0$ as $n \to \infty$.

This means,

$$\inf\left\{c > 0 : L\int_{\mathcal{B}} \theta\Big(t, (\xi_n - \xi)(t)\Big)d\mu_{\infty} \leq 1\right\} \to 0$$

$$\implies L\int_{\mathcal{B}} \theta\big(t, (\xi_n - \xi)(t)\big) \to 0$$

$$i.e., \ \theta(t, (\xi_n - \xi)(t) = 0 \ \mu_{\infty} - a.e. \ as \ n \to \infty$$

$$\xi_n - \xi = 0 \ a.e. \ as \ n \to \infty.$$

As, $\xi_n \in \mathcal{L}^{\theta}(X, \mu_{\infty})$, this implies $\xi_n \in \mathcal{H}^{\theta}(\mathcal{X}, \mu_{\infty})$ and $\xi_n - \xi = 0$ as $\mu_{\infty} - a.e.$ Now using Theorem 10.2,

$$HK\int_{\mathcal{B}} (\xi_n - \xi)d\mu_{\infty} = 0.$$

Lastly, from Definition 10.1 (d)

$$HK\int_{\mathcal{B}} \theta(t, (\xi_n - \xi)(t)) = 0.$$

So,

$$\inf\left\{c > 0 : HK\int_{\mathcal{B}} \theta(t, (\xi_n - \xi)(t))d\mu_{\infty} \leq 1\right\} \to 0.$$

Hence, $||\xi_n - \xi||_{(\mathcal{X},\mu_{\infty})} \to 0$ as $n \to \infty$. □

Theorem 10.12 *Modular convergence in $\mathcal{L}^{\theta}(\mathcal{X}, \mu_{\infty})$ is norm convergence in $\mathcal{H}^{\theta}(\mathcal{X}, \mu_{\infty})$.*

Proof Let $(\xi_n) \in \mathcal{L}^{\theta}(\mathcal{X}, \mu_{\infty})$ be modular convergent to $\xi \in \mathcal{L}^{\theta}(\mathcal{X}, \mu_{\infty})$. Then for $c > 0$ (depending on the sequence of function (ξ_n)) such that $\rho\big(c(\xi_n - \xi)\big) \to 0$ as $n \to \infty$. That is,

$$L \int_{\mathcal{B}} \theta(t, c(\xi_n - \xi)) d\mu_{\infty} \to 0 \; as \; n \to \infty$$

$$i.e., \; \lim_{n\to\infty} L \int_{\mathcal{B}} \theta(t, c(\xi_n - \xi)) d\mu_{\infty} = 0$$

$$i.e., \; \theta(t, c(\xi_n - \xi)) = 0 \; \mu_{\infty} \; a.e. \text{ when } n \to \infty.$$

For $c > 0$,

$$\inf\left\{\frac{1}{c} > 0 : HK \int_{\mathcal{B}} \theta\left(t, \frac{\xi_n - \xi}{\frac{1}{c}}\right) d\mu_{\infty} \leq 1\right\} \to 0 \; \mu_{\infty} - a.e.$$

as $n \to \infty$. So $||\xi_n - \xi||_{(\mathcal{X}, \mu_{\infty})} \to 0$ as $n \to \infty$. This completes the proof. □

References

1. Alexiewicz, A.: Linear functionals on Denjoy-integrable functions. Collectanea Mathematica **1**, 289–293 (1948)
2. Bartle, R.G., Sherbert, D.R.: Introduction to Real Analysis 3rd edn. Wiley-India (2010)
3. Fernandez, A., Mayoral, F., Naranjo, F., Paual, P.J.: Weakly sequentially complete Freáchet spaces of integrable functions. Archiv der Mathematik **71**(3), 223–228 (1998)
4. Hazarika, B., Kalita, H.: Henstock-Orlicz spaces and its dense spaces. Asian Eur. J. Math. **14**(7), 1–17 (2021)
5. Kalita, H., Hazarika, B.: Modular convergence in H-Orlicz spaces of Banach valued functions. Rendiconti del Circolo Matematico di Palermo **72**(8), 3905–3916 (2023)
6. Kaminska, A., Hudzik, H.: Some remarks on convergence in Orlicz space. Annales Societatis Mathematicae Polonae, Series I: Commentationes Mathematicae, XXI (1979)
7. Lindenstrauss, J., Tzafriri, L.: Classical Banach Spaces II, Function Spaces. A Series of Modern Surveys in Mathematics. Springer, Berlin-Heidelberg-New York (1979)
8. Maligranda, L.: Orlicz spaces and Interpolation, Seminarios De Matematica, Departmento de Matematica, Universidade Estadual de Campinas (1989)
9. Musielak, J., Orlicz, W.: On modular spaces. Studia Mathematica **18**, 49–65 (1959)
10. Skaff, M.S.: Vector valued Orlicz spaces, II. Pac. J. Math. **28**(2), 1–18 (1969)

Index

B. Hazarika, H. Kalita, *Theory of Henstock–Orlicz Spaces*, Mathematical Marvels: Texts and Monographs in the Spirit of CR Rao,
https://doi.org/10.1007/978-981-96-9548-5

MIX
Papier aus verantwortungsvollen Quellen
Paper from responsible sources
FSC® C105338

If you have any concerns about our products,
you can contact us on
ProductSafety@springernature.com

In case Publisher is established outside the EU,
the EU authorized representative is:
Springer Nature Customer Service Center GmbH
Europaplatz 3, 69115 Heidelberg, Germany

Printed by Libri Plureos GmbH
in Hamburg, Germany